environment /ənˈvaɪrə(n)m(ə)nt/

noun

• the area surrounding a place or thing; the environs, surroundings, or physical context: *Bairuth, with its kind picturesque environment.*

• the physical surroundings or conditions in which a person or other organism lives, develops, etc., or in which a thing exists; the external conditions in general affecting the life, existence, or properties of an organism or object: *the organism is continually adapted to its environment*.

• the natural world or physical surroundings in general, either as a whole or within a particular geographical area, esp. as affected by human activity: *the situation is clouded by a widespread confidence that this impact of man upon environment can continue indefinitely.*

Oxford English Dictionary, 3rd Edition

IN THIS ISSUE

LA+ ENVIRONMENT
EDITORIAL

The term "environment" is so common, and its meaning often taken for granted, that it is easy to forget how recently it has proliferated. To provide context, this issue begins with an essay by environmental historian Etienne Benson, who explores the history of the concept and its ever-evolving assemblage of ideas and practices. Following on the many "environmentalisms" discussed in Benson's opening essay, historian Chad Montrie challenges the conventional narrative that ties the rise of environmentalism in the US to figures like Rachel Carson and Ian McHarg. As Montrie argues, this view overlooks the longer history of activism led by working-class individuals, labor unions, and civil rights groups. Of course, these efforts are not just a thing of the past, as the conversation between Jessica Varner and Jess Conard shows. Conard, who serves as the Appalachia director of Beyond Plastics, became politically active after the train derailment and vinyl chloride spill in East Palestine, Ohio, in 2023. We also spoke with policy expert Maria Ivanova about the development of a global treaty on plastics pollution, the work of the UN Environment Programme, and the difficulties of global environmental governance. As we go to press, US environmental regulations designed to protect people, air, and water are once again being rolled back, and efforts to address climate change are being abandoned. While regulations do not always guarantee enforcement–as illustrated in the conversation between Varner and Conard–they provide a legal framework to hold those responsible for environmental harm accountable. Despite the rapid gutting of environmental programs and a weakened regulatory environment on pollution, this set of essays and interviews serves as a hopeful reminder of the ongoing efforts by many individuals who continue to press for change, even in the face of setbacks and adversity.

The next grouping of essays revisits mid-20th-century examples where landscape design and art are employed to raise awareness of or mitigate the effects of various environmental problems. Art historian Melissa Ragain writes about artist Herbert Bayer's large-scale installations created for ARCO petroleum company from the mid-1960s to the mid-1980s. Landscape historian Sonja Dümpelmann explores the study of plant acoustics from the 1950s to the 1970s, which grew out of concerns about noise pollution arising from rapid growth in private automobiles and air traffic in the post-WWII era. In their essay, landscape architects Marcus Owens and Christina Antiporda describe how the development of nuclear weapons provided a unifying view of the environment, characterized as a "technological sublime" that evoked both awe and terror. They examine various efforts to communicate danger to Earth's future inhabitants through signage and landscape design. Architectural historian Rami Kanafani brings to light an interesting case study on interdisciplinarity from the 1970s and '80s, where New Age spiritualists collaborated with philosophers, designers, and scientists in response to the growing awareness of environmental degradation.

The next series of interviews and essays examines the difficulty of fostering linkages between local and global efforts to address the climate crisis. We interviewed activist-scientist Michael Oppenheimer, who has been involved with climate policy for over four decades. Social scientist Holly Jean Buck decries the lack of scalar imagination regarding the energy transition, arguing that this is a "failure to build." Her essay calls on designers to take up this challenge in more systemic ways. We spoke with historian Dipesh Chakrabarty about how the concept of the planetary differs from that of the global environment and the philosophical challenges of historicizing our human condition within a planetary context. Also underscoring the importance of environmental history in the context of the Anthropocene, environmental historian Vinita Damodaran emphasizes the need to carefully analyze how local communities and national governments have historically responded to environmental disruptions. This understanding can help us formulate more equitable approaches to addressing the climate crisis.

The last four essays illuminate our troubled relationship with nature, including efforts to tame it with various technologies and representations. Historians and philosophers of science, Jip van Besouw and Lino Camprubí describe the gradual transformation of seas from being viewed as unchangeable and beyond human influence to being increasingly territorialized and thus akin to *terra*. Architect Andrew Witt discusses the immersive effects of 19th- and early 20th-century panorama drawings of distant landscapes, using drawings from the 1910 *Terra Nova* Antarctic expedition to reflect on the changing concept of the environment. Architect Diana Guo begins her essay by considering the evolving technologies of environmental "preservation," from nature dioramas found in museums of natural history to commercial-corporate efforts to create virtual reality environments comprising real and imaginary wildlife to recent attempts at resurrecting long-extinct species using DNA extraction technologies. In the final essay, historian and ecologist Laura J. Martin also questions the notion of preservation and urges us to rethink the concept of "habitat." While this concept is foundational to ecology, Martin argues that it does not help us find new ways to coexist with the diverse forms of life that share our planet.

Karen M'Closkey
Issue Editor

ENVIRONMENTS, ENVIRONMENTALISMS,
ENVIROTECHNIQUES, ENVIRONMENTS, EN
VIRONMENTALISMS, ENVIROTECHNIQUES,
ENVIRONMENTS, ENVIRONMENTALISMS,

ETIENNE BENSON

ENVIROTECHNIQUES, ENVIRONMENTS, EN
VIRONMENTALISMS, ENVIROTECHNIQUES,
ENVIRONMENTS, ENVIRONMENTS, ENVIRO
NMENTALISMS, ENVIROTECHNIQUES, ENV
IRONMENTS, ENVIRONMENTALISMS, EN
VIROTECHNIQUES, ENVIRONMENTS, ENVI
RONMENTALISMS, ENVIROTECHNIQUES,
ENVIROTECHNIQUES, ENVIRONMENTS, EN

Etienne Benson is a historian of environmentalism and the environmental sciences. Since 2022, he has been a director at the Max Planck Institute for the History of Science, Berlin, where he heads the Department on Knowledge Systems and Collective Life. His most recent book is *Surroundings: A History of Environments and Environmentalisms* (Johns Hopkins University Press, 2020). He is currently working on a history of geomorphology, the science of landforms.

+ ENVIRONMENTAL HISTORY, HISTORY OF SCIENCE

In 1805, André Thouin, a professor at the Museum of Natural History in Paris, published an account of the various "instruments of modern invention" in use in the Museum's botanical garden.[1] Such instruments were necessary, Thouin explained, because the garden was home to plants of a tremendous variety, each suited to its own native climate and soil: plants from frigid Kamchatka and from torrid Africa, alpine plants and aquatic plants, trees requiring dense, well-established soils, and trees requiring soils that were light and full of newly decomposed vegetable matter.

Founded in 1793 in the wake of the Revolution, the Museum of Natural History was tightly tied to the expansionary aims of Napoleonic France, even if such aims remained implicit in Thouin's article and most other reports in the Museum's *Annales*.[2] The "modern inventions" described by Thouin were tools of empire, designed both to explain and to exploit the plants gathered from throughout the world by French merchants, sailors, colonial officials, and naturalists. These inventions included *parapluies* to protect plants from the rain, *contresols* to protect them from the sun, *châssis portatifs* to protect them from cold winters, and *grillages* to protect them from birds and other pests.

Despite Thouin's insistence on their status as modern inventions, these umbrellas, shades, frames, and fences were hardly revolutionary in themselves. On the contrary, they built on the work of numerous gardeners and botanists who had maintained the garden's living collection since its establishment in the middle of the 17th century, and they in turn drew on the practical expertise of many others throughout France and beyond.[3] It required no stroke of genius or esoteric program of study to recognize that human-built structures, from tiny flower pots to gargantuan greenhouses, could help to deliver the right amounts of sun, wind, and water to plants according to their specific needs.

What *was* novel, perhaps even revolutionary, was the way these devices were deployed in a particular material and social setting to give embodied reality to an emerging set of ideas about the form and function of living things in relation to their surroundings – ideas that were being developed at the time by Georges Cuvier, Jean-Baptiste Lamarck, and other naturalists at the museum who believed that the organization of living bodies could only be explained in reference to their material surroundings and ways of life. It was not the devices themselves that were worthy of note, in other words, but rather their use as part of what one might call *envirotechniques*: methods of transforming environmental ideas into material realities and vice versa.[4]

The concept of "environment" is neither a universal concept to be found in all human societies at all times, nor is it the invention of 20th-century environmentalists.[5] Rather, it is a shifting assemblage of ideas, practices, and technologies that was first developed in certain European institutions of science around the end of the 18th century–the Museum of Natural History in Paris among them–and that has been spreading and diversifying into a bewildering variety of environments, environmentalisms, and envirotechniques ever since.

At first glance, the idea that the concept of environment is universal, or at least ancient, might seem more compelling than the idea that it was invented a few hundred years ago. After all, haven't people always thought about their surroundings? In the history of Western medicine, just to take one example, there is a long tradition of using factors that we would now readily describe as "environmental" to explain disease. The Hippocratic tradition, for example, links classical antiquity's concerns with "airs, waters, and places" to new forms of medical theory and practice that emerged in the early modern period, which were designed to categorize and to influence connections between external conditions–enervating climates, morbific miasmas, and so forth–and the body's own humoral balance.[6]

Indeed, far from seeing environmental thought as a modern phenomenon (whether one dates "modern" to the late 18th century or the mid-20th century), some historians have looked to neo-Hippocratic medicine as a resource for mounting a critique of biomedicine from within the Western medical tradition.[7] They have argued that what is actually modern–and aberrant, from a long-term historical perspective–is the shift of attention away from the surroundings and toward internal causes and specific agents, such as the pathogenic microbes discovered by the likes of Robert Koch and Louis Pasteur. Environmental thought is, in this view, profoundly nonmodern in its emphasis on the porosity of the body and its co-constitutive entanglement with the world around it.

Seen through this lens, the case for the continuity and nonmodernity of environmental thought seems indisputable. But if the significance of a concept cannot be captured through abstract definitions or philosophical analyses but rather must be understood through its embodiment in particular instruments and practices in specific material and social settings–that is, through envirotechniques–then the arguments for continuity begin to seem less convincing. It may well be that people have always been concerned with their surroundings, but that does not mean that they have always been concerned in the same way, or that the differences among their concerns can be dismissed as superficial or inconsequential. On the contrary, assuming that the modern concept of environment applies to all times and places may be misleading.

In the case of environmental medicine, then, it is far less interesting to note that European physicians in the 19th century were interested in the effects of the surroundings on health, just as many before them had been, than to ask what specific (enviro)techniques they deployed to put their own distinctive ideas about those effects into practice. *That* question cannot be answered in the abstract; it requires accounting for, among other things, the development of new methods for quantifying geographical variations in health and disease, patterns of migration linked to European colonialism and global capitalism, conceptions of biological race as a determinative factor for health, laboratory procedures for investigating microscopic organisms, and sanitary techniques for controlling human waste and regulating the movement of disease vectors.

What resulted from this welter of developments was not the first recognition of the influence of the surroundings on health and disease but rather a distinctively modern form of that recognition – one that, in the last quarter of the 19th century, was gradually coming to be associated with a new word: environment.

In the English language, the word environment has a relatively short history, at least in comparison to some of the terms most closely related to it, such as "nature," "climate," "land," or "earth." While environment can be found as early as the 13th century, it is rare before the 19th century, and when it appears, its meaning is usually quite different than in modern usage, referring to the act of surrounding rather than to the surroundings themselves. The *Oxford English Dictionary* provides an example of such usage as late as 1888, when an author for *Century Magazine*, describing a naval battle, wrote: "A similar blow on the other side, or a close environment of the heavy guns of the other ships, would have captured or sunk the ram."[8]

By the 1880s, this older usage, never particularly common, was already being overshadowed by a new usage that had made its first tentative appearance in the 1820s, gained firmer purchase in the 1850s, and expanded rapidly thereafter. In this new usage, environment was not an act but rather an *object* of a very particular sort – namely, the object comprising all the surrounding circumstances that were relevant to the life of an individual or organism.

This modern usage was first introduced in the 1820s by Thomas Carlyle, who used it to explain the works and lives of artists and historical figures, particularly what he saw as heroic figures associated with Romanticism and with the French Revolution. But in this form it remained an awkward outlier, largely confined to Carlyle's own texts, with their distinctive and polarizing style. One correspondent of his in the 1830s even included the word in a list of "positively barbarous" neologisms that, in his view, marred Carlyle's otherwise admirable literary contributions.[9]

Environment broke free of these narrow confines, so to speak, in the 1850s, when it featured prominently in attempts to develop grand syntheses that could encompass all the emerging scientific specializations of the era. It was not an accident that this new sense of environment emerged just as the term "scientist," first proposed in 1833, was beginning to gain wider acceptance.[10] Both neologisms reflected an attempt to grapple with the diversification of scientific disciplines that was rapidly rendering obsolete such terms as "natural philosopher" and "natural historian" and raising questions about what, if anything, these new disciplines–from physics to physiology–had in common.

The first significant text that used environment in this broader context was Harriet Martineau's condensed English translation of Auguste Comte's *Cours de philosophie positive*, which appeared in 1853 and rapidly became the standard reference on Comte's thought in the English language.[11] Martineau turned to the then-unfamiliar term environment to translate Comte's *milieu*, a concept he had grounded in his understanding of the biology of the day–including physiological research conducted at the Museum of Natural History–but had also transformed and expanded for his own purposes.[12]

For Comte, the concept of *milieu* helped to unify the dizzying proliferation of sciences that had taken place since the late 18th century. Everything from the action of the tiniest animalcule to the progress of human civilizations, he argued, could be explained through the reciprocal relation between the internal organization of the organism (or community, society, or civilization) and its external *milieu*. Organism and *milieu* were an inseparable pair, each determining the other. As Martineau expressed it in her translation, "If the idea of life is really inseparable from that of organization, neither can be severed, as we have seen, from that of a medium or environment, in a determinate relationship with them."[13]

Herbert Spencer, the writer who was to become most closely associated with the term environment in subsequent decades, most likely encountered the term in this broadened sense in Martineau's text. His own *Principles of Psychology*, published two years later in 1855, joined Martineau's translation in introducing the term environment to an English-speaking audience, and Spencer repeatedly returned to the concept in his lifelong attempt to develop a "synthetic philosophy." Like Comte (and Martineau), he was deeply convinced that the relationship between organisms and their environment was critical both to understanding life and to building bridges across scientific disciplines.[14]

In the late 19th century, as industrialization transformed city and countryside alike and European empires expanded their reach, more and more European scientists and social thinkers became concerned with the effects of the material surroundings on individuals, communities, and societies, both at home and abroad. As early as the 1880s, even those dubious

1 André Thouin, "Description et usage de plusieurs ustensiles de moderne invention, propres à la culture d'un grand nombre de plantes dans les écoles de botanique," *Annales du Muséum d'histoire naturelle* 6 (1805): 236–52.

2 E.C. Spary, *Utopia's Garden: French Natural History from Old Regime to Revolution* (University of Chicago Press, 2000).

3 Ibid., 49–98.

4 Cf. Michelle Murphy, *Sick Building Syndrome and the Problem of Uncertainty: Environmental Politics, Technoscience, and Women Workers* (Duke University Press, 2006); John Tresch, "Cosmologies Materialized: History of Science and History of Ideas," in Darrin M. McMahon & Samuel Moyn (eds), *Rethinking Modern European Intellectual History* (Oxford University Press, 2014), 153–72; Sara B. Pritchard & Carl A. Zimring, *Technology and the Environment in History* (Johns Hopkins University Press, 2020); Sverker Sörlin & Nina Wormbs, "Environing Technologies: A Theory of Making Environment," *History and Technology* 34, no. 2 (2018): 101–25; Adam Wickberg & Johan Gärdebo (eds), *Environing Media* (Routledge, 2023).

5 Paul Warde, Libby Robin & Sverker Sörlin, *The Environment: A History of the Idea* (Johns Hopkins University Press, 2019).

6 Conevery Bolton Valencius, "Histories of Medical Geography," in Nicolaas Rupke (ed.), *Medical Geography in Historical Perspective*, Suppl. No. 20 of *Medical History* (Cambridge University Press, 2000), 3–28; Christopher Sellers, "Health, Work, and Environment: A Hippocratic Turn in Medical History," in Mark Jackson (ed.), *The Oxford Handbook of the History of Medicine* (Oxford University Press, 2012), 450–68.

7 E.g., Gregg Mitman, *Breathing Space: How Allergies Shape Our Lives and Landscape* (Yale University Press, 2007); Linda Nash, *Inescapable Ecologies: A History of Environment, Disease, and Knowledge* (University of California Press, 2007).

8 Oxford English Dictionary, online, entry for *environment (n.)*: https://www.oed.com/dictionary/environment_n (accessed January 26, 2025).

9 Ralph Jessop, "Coinage of the Term Environment: A Word Without Authority and Carlyle's Displacement of the Mechanical Metaphor," *Literature Compass* 9, no. 11 (2012): 708–20.

10 Steven Shapin, *The Scientific Life: A Moral History of a Late Modern Vocation* (University of Chicago Press, 2009), 322n7.

11 Harriet Martineau, *The Positive Philosophy of Auguste Comte*, Vol. I (John Chapman, 1853).

12 John Tresch, *The Romantic Machine: Utopian Science and Technology After Napoleon* (University of Chicago Press, 2012), 269–75.

13 Martineau, *Positive Philosophy*, 365.

14 Trevor Pearce, "From 'Circumstances' to 'Environment': Herbert Spencer and the Origins of the Idea of Organism–Environment Interaction," *Studies in History and Philosophy of Biological and Biomedical Sciences* 41, no. 3 (2010): 241–52.

of Comte's and Spencer's grand syntheses were beginning to use the concept of environment–and the word–to explain the behavior of microbes, galaxies, engines, societies, and much else besides.

At the end of the day, a word is just a word, with little power to change the world – unless, that is, it comes to be embodied in a set of practices and instruments that are effective within a particular social and material setting.

Over the course of the 19th century, a number of envirotechniques were developed that made the term environment into something more than an intriguing neologism or even a compelling idea. When people spoke or wrote about environment, they usually had one or more of these specific envirotechniques in mind. Hidden under the seeming uniformity of the term, in other words, was a wild proliferation of closely related concepts, none of them exactly identical, none of them entirely reducible to another.

For the sanitary engineer who drained marshes to reduce populations of mosquitoes that spread yellow fever and other diseases in the Caribbean, there was one kind of environment; for the plant breeder who sought to maximize the productivity of a crop by adapting it to local climates and soils, there was another; for the labor organizer who sought to reduce injuries in the factory by advocating for better lighting and safer materials, there was yet another. And so on, ad infinitum. It was precisely because the concept of environment could be adapted to describe a wide range of techniques for representing and intervening in the world that its rise in the late 19th century was so meteoric.

Still, even if the concepts that fit under the umbrella of environment were diverse, they also shared certain family traits. This was not because there was (or is) some essential core to the concept of environment but because, like other kinds of families, each new addition inherited some things from its predecessors while altering or abandoning others, without changing so much that the family resemblance was lost entirely. The unity of modern environmental concepts, such as it is, emerges from the situated histories of ideas and practices that both continue and transform what came before, among communities of speakers and doers with some commitment to collective life and mutual intelligibility.

From the 19th century to the present, those who have adopted environment as a keyword and who have given it reality by embodying it in practices and instruments–those who have developed and deployed envirotechniques, in other words–have generally shared certain convictions. They include the convictions that explanations of natural phenomena should be grounded in material entities and forces (however "material" may be defined), that the most authoritative explanations come from people with scientific credentials and training (however "scientific" may be defined), and that life can best be explained in terms of relationships between organized bodies and their surroundings, each clearly distinguishable from the other (however "life" may be defined).

As the parentheticals above suggest, even within the bounds of these shared premises, there is a great deal of room for variation, including challenges to widely shared assumptions about the nature of matter, knowledge, and life. Nowhere is the tension between commonality and difference more evident than in the modern environmental movement, whose rhetoric has often advanced a universalizing understanding of *the* environment but whose practices reveal a much messier and more contested world of environments and environmentalisms in the plural.

This situated messiness is visible even in the most canonical of US environmental texts, Rachel Carson's *Silent Spring*.[15] While it was rooted in the long tradition of environmental thought–some of its rhetoric would not have seemed out of place in the late 19th century–it was neither a straightforward continuation of Comtean or Spencerian environmentalism nor an elaboration on the themes of the landmark environmental texts of the immediate postwar period, William Vogt's *Road to Survival* and Fairfield Osborn's *Our Plundered Planet*, both published in 1948.[16] *Silent Spring* proposed an environment, and an environmentalism, that was all its own.

Whereas late-19th-century environmental thinkers had focused on the adaptation of organisms to their surroundings, and postwar neo-Malthusians such as Vogt and Osborn were preoccupied with the balance between resources and populations, Carson crafted her own distinctive blend of toxicology, ecology, genetics, and evolutionary theory to explain why human-made "biocides" of various kinds posed an existential threat to life on earth. In the years following *Silent Spring*, the emerging environmental movement welded this scientific argument to certain methods of risk assessment, strategies of mass mobilization, and forms of state action – collectively, a set of envirotechniques that materialized a particular conceptualization of the environment.

There was not one postwar environmentalism or one postwar conceptualization of the environment, in other words, but many. Each alternative environmentalism assembled its own practices and methods to reconfigure both the environment and the environed subject, subtly or not-so-subtly changing the scales of analysis, metrics of harm or quality, and modes of intervention. The environmental justice movement of the 1980s and 1990s, for example, deployed new methods of "popular epidemiology" and public protest to push back against mainstream environmentalism's race- and class-agnostic universalism, while the climate-change activism of the 1990s and 2000s expanded the spatial and temporal scope of the environment with the help of computational models and novel political-scientific institutions such as the IPCC.[17]

As these examples suggest, environmentalism was (and is) a broad church in which many variations on the concept of environment, each embodied through a particular set of envirotechniques and developed in a particular social and material setting, have jostled for legitimacy, attention, and resources. Some of these variations, including many forms of official environmentalism, have taken the concept's affinities toward materialism, scientism, and individualism to reductive extremes – constructing environments and environed subjects for which only physical form and function are allowed to matter, for which dominant science is the only acceptable source of expertise, and for which the border between individuals and their surroundings is distinct and fixed. These reductions and exclusions are often coupled with a narrow view of the issues that can be counted as "truly environmental."[18]

But it would be a mistake to conclude that these rigid environmentalisms, however dominant they may be, exhaust the possibilities of the concept of environment – just as it would be a mistake to think that the myriad environments, environmentalisms, and envirotechniques that have emerged over the past two centuries are all, in their essence, reducible to the ideas and methods that Thouin and his colleagues at the Museum of Natural History developed in the decades around 1800. On the contrary, like any historical process, the development of the concept of environment has featured constant reversals, rebellions, unravelings, reweavings, retoolings, and reinventions – all of which have produced a past (and present) that is much more complex, interesting, and perhaps even hopeful than the trajectory of a single, fixed idea through time could ever be.

15 Rachel Carson, *Silent Spring* (Houghton Mifflin, 1962).

16 William Vogt, *Road to Survival* (William Sloane, 1948); Fairfield Osborn, *Our Plundered Planet* (Little, Brown and Company, 1948). See Warde, Robin & Sörlin, *Environment*, 6–24.

17 Dorceta E. Taylor, *Toxic Communities: Environmental Racism, Industrial Pollution, and Residential Mobility* (NYU Press, 2014); Phil Brown, "Popular Epidemiology and Toxic Waste Contamination: Lay and Professional Ways of Knowing," *Journal of Health and Social Behavior* 33, no. 3 (1992): 267–81; Paul N. Edwards, *A Vast Machine: Computer Models, Climate Data, and the Politics of Global Warming* (MIT Press, 2010).

18 For conceptual critiques of environment along these lines, see, e.g., Donna J. Haraway, *Staying with the Trouble: Making Kin in the Chthulucene* (Duke University Press, 2016); Bruno Latour, *Facing Gaia: Eight Lectures on the New Climatic Regime*, trans. Catherine Porter (Polity Press, 2017).

CHAD MONTRIE

THE OTHER SILENCE IN SILENT SPRING

Chad Montrie is a professor in the history department at the University of Massachusetts, Lowell. He has authored several books, including *The Myth of Silent Spring: Rethinking the Origins of American Environmentalism* (2018). From 2022–2023, he was a Fulbright Canada Research Chair at the University of Calgary, where he completed research for "'What is Labour's Stake?': Workers and the History of Environmentalism in Alberta," *Labour/Le Travail* (2024), and "'Agenda for the 1970s': A Genealogy of Organized Labour's Environmental Activism in Ontario," *Papers in Canadian History and Environment* (2024).

+ ENVIRONMENTAL JUSTICE, HISTORY

Following: Walter Reuther and UAW aid in protests with César Chávez during the Delano grape strike in 1965.

Published to wide acclaim in 1962, Rachel Carson's book *Silent Spring* was a deftly written exposé of synthetic pesticides. The "silence" the title alluded to was what might happen with the continued unregulated use of those dangerous chemicals, and chapter by chapter, the popular science writer methodically evoked the possibility of an apocalyptic world where life could not persist. Serialized in the *New Yorker* in 1961 and then featured in a television program in 1963, *Silent Spring* introduced millions of Americans to the principles of ecology. It turned their attention toward a steadily worsening environmental and human health threat. For doing that, countless observers credited Carson's book with catalyzing the American environmental movement, both at the time and since its original publication. In a jacket endorsement for the first edition, US Supreme Court justice and conservationist William O. Douglas called *Silent Spring* "the most revolutionary book since *Uncle Tom's Cabin*," referring to the "big book" origin story for the American Civil War. Famed children's author and essayist E. B. White also described it in a review as "an *Uncle Tom's Cabin* of a book,–the sort that will help turn the tide."[1]

There was another "silence" in *Silent Spring*, however, an absence its admirers overlooked yet one that had, and has, profound implications for understanding environmental problems as well as confronting and solving them. Entirely missing from the much-lauded jeremiad was any awareness of how class, race, and ethnicity determined people's exposure to environmental hazards, who was responsible for those hazards, and what someone could do about them. While registering a compelling indictment against the unfettered use of pesticides, the book made almost no mention of farm workers, even though field and orchard hands had some of the most direct, frequent, and prolonged contact with the chemicals. Moreover, the few times *Silent Spring* hinted at the agricultural laborers' plight, the references were brusque generalizations. In chapter 3, "Elixirs of Death," Carson briefly noted the chance of exposure from dusting and spraying on farms, for instance, but the following and only full-sentence acknowledgment of the risks farm workers faced was not until chapter 12, "The Human Price." "The sudden illness or death of farmers, spraymen, pilots, and others exposed to appreciable quantities of pesticides are tragic," she wrote, "and [they] should not occur."[2]

Like George Perkins Marsh's classic *Man and Nature*, published nearly one hundred years before, Carson's *Silent Spring* told a story about a universal "man." Earlier working titles for the book had included "Man Against Nature" as well as "Control of Nature," the latter suggesting that the culprit for environmental threats was so obviously "man" that he need not be named. The phrase "Silent Spring" did that, too, besides implying the most noteworthy impact of "man's war against nature" was poisoning songbirds. The final title and the book itself neglected to place specific blame on those who had the means and power to wreak the greatest harm and ignored how social divisions and inequality condemned whole groups of people to a disproportionate share of it. To be sure, in the 1963 CBS television program, "The Silent Spring of Rachel Carson," the opening shots were clear about the danger pesticides posed to songbirds and people alike, although then the concern was focused on those living in the suburbs. The program began with Carson on her porch at her seaside home in Maine, along with her voice-over, reading a line about chemicals being the sinister partners of radiation and nonselective weapons that kill everything, a warning paired with footage of a crop duster flying over tract houses.[3]

Similarly, Carson's contemporary in landscape architecture, Ian McHarg, was coming to terms with ecology and its implications for his profession, and his influential work was burdened by an abstract "man" (or undifferentiated "we") too. McHarg first developed his ideas at the University of Pennsylvania in a course titled "Man and Environment" and published them in a 1963 essay by the same name, in between which he became a public personality by hosting a high-profile CBS television show called "The House We Live In."[4] "The natural scientist states that no species can exist without an environment," McHarg explained in the essay, "Man must learn this prime ecological lesson of interdependence. He must see himself linked as a living organism to all living and all preceding life." A few years later, in 1969, on the eve of the first Earth Day, McHarg expanded on that point in a book, *Design with Nature*. "In the quest for survival, success and fulfillment," he mused there, "the ecological view offers an invaluable insight. It shows the way for the man who would be the enzyme of the biosphere—its steward, enhancing the creative fit of man-environment."[5]

In the "Foreword" to *Design with Nature*, Lewis Mumford had allowed he could not predict the book's fate, "but on its intrinsic merits I would put it on the same shelf that contains as yet only a handful of works in a similar vein, beginning with Hippocrates, and including such essential classics as those of Henry Thoreau, George Perkins Marsh, Patrick Geddes, Carl Sauer, Benton MacKaye, and Rachel Carson." As it turned out, McHarg's work was (and is) much praised and respected despite an astoundingly myopic (or at least overly simplistic) view of urban demographics. Most notably, perhaps, *Design with Nature*'s final complete chapter attempted to demonstrate the promise and potential of McHarg's now ubiquitous "layer-cake" model by looking at so-called "urban pathology" in Philadelphia. It managed to do this without using the word "racism" and without acknowledging the accelerated urban disinvestment caused by White flight to suburbs and the racial exclusion residents subsequently conducted there. "The pattern is very clear," McHarg explained, "the heart of the city is the heart of pathology and there is a great concentration of all types of pathology encircling it." While it was "premature to predict correlations," he concluded, "the single obvious one is not poverty, but density," employing a word and concept that was at the very least a strange euphemism and, on its own, was no explanation at all.[6]

Unquestionably, both Carson and McHarg made unique and significant contributions to the spread of ecological thinking and environmental engagement in America. But a "big book" origin story for the environmental movement—whether citing *Silent Spring* or *Design with Nature*—erroneously elevates the work of the lone scientist or individual landscape architect in making environmentalism happen. That interpretation misses how the movement actually took root and spread, which was mainly because many unheralded, ordinary people contemplated, worried over, and acted to address environmental problems. What's more, they usually did this not because they read a particular book (or watched a specific television program), but because they were driven to it by their own perilous circumstances. In some cases, this was years, if not decades, before *Silent Spring* (or *Design with Nature*) was published or, in other instances, with little or no evidence that they were very influential. Likewise, peoples' circumstances were typically interlayered with economic exploitation, structural racism, and ethnic prejudice—salient social dimensions that *Silent Spring* and *Design with Nature* and their enthusiasts missed—and, more often than not, people reacted and struggled accordingly, simultaneously incorporating or manifesting their environmental concerns as part of labor, civil rights, and other movements.[7]

"A Massive Mobilization"

Contrary to the notion that *Silent Spring* started the American environmental movement, working people were already concerned about various environmental problems in the early 19th century. This concern followed the onset of industrialization and consequent urbanization, and it carried over into the 20th century, as manufacturing spread and cities grew, giving rise to ever-worsening working and living conditions. In the post-World War II era, the most significant manifestation of organized working-class environmental consciousness and activism (counter-intuitive to some) was the United Auto Workers (UAW). That was due in no small part to the union's progressive leadership, namely its president, Walter Reuther, a Cold War liberal who fostered a broad social unionism that addressed workers' outdoor recreation as well as occupational health and safety hazards and air and water pollution in the larger community.

The UAW established a Recreation Department in 1939, and during the 1940s that department joined with Michigan's many county-level working-class hunting and fishing clubs to advocate for sportsmen's interests, notably those that most benefited workers, such as keeping permit fees low and allowing hunting and fishing on Sundays. The county clubs and "international" union also began lobbying for state legislation to control municipal and industrial waste in waterways–achieving their first success in 1947–and in the 1950s, the UAW began to couple that with demands for federal action. Subsequently, in 1965, the union hosted a conference on clean water at Solidarity House in Detroit, drawing more than one thousand union, sportsmen, and environmental delegates, from both the United States and Canada, the largest and most diverse meeting of the kind to date. "I am hopeful that this conference here today can be the beginning of a massive mobilization of a great citizen crusade," Walter Reuther declared in his opening speech, "not only in terms of fighting for clean water, but for cleaning up the atmosphere, and the highways, and the junkyards and the slums and all other things that we need to do to create a total living environment worthy of free men."[8]

Two years later, in 1967, the UAW established a Department of Conservation and Resource Development. The union did this, Director Olga Madar explained, "because our members and their families are directly affected by the environment around them, both inside and outside of the plants in which they work." Yet the UAW was not concerned exclusively with its own membership. "Air and water pollution, the desecration of the land, and the unwise use of natural resources," Madar said, "are of great concern to us all." True to that sentiment, the union helped fund Environmental Action, the group coordinating the first Earth Day in April 1970, and coinciding with that momentous event, delegates at the UAW's 22nd constitutional convention in Atlantic City called for an "environmental bill of rights." In the summer (shortly after Reuther's tragic death in a plane crash), the UAW also hosted a varied group of participants for a conference on "urban and industrial pollution" at its Black Lake retreat center. And, in 1976, it organized another Black

Lake meeting, "Working for Environmental and Economic Justice and Jobs," where union President Leonard Woodcock declared a "common cause between union members and environmentalists–between workers, poor people, minorities, and those seeking to protect our natural resources."[9]

In line with their broad-minded principles, the UAW aided other workers attempting to organize as well, including migrant Mexican and Mexican American farm workers in Southern California, where capitalism had come to the countryside to make "factories in the fields." Community activists César Chávez and Dolores Huerta arrived in the San Joaquin Valley in 1962, and together, they helped field and orchard hands establish the United Farm Workers Organizing Committee. The workers' primary grievances were not only meager wages, long hours, and inadequate housing but also makeshift or nonexistent protection from exposure to pesticides and herbicides. Likewise, since the campaign coincided with the rise of a Chicano empowerment movement, they linked their health concerns from chemical exposure to a pattern of racial discrimination, a critical connection other activists would later term "environmental racism." In fact, by employing an expansive understanding of the problems they faced and a range of militant tactics, the farmworkers were also effectively inventing "environmental justice" activism, several decades before the movement was supposedly established.[10]

In 1965, following investigations by their own health and safety commission, the renamed United Farm Workers called for a consumer boycott of table grapes, and as part of that campaign, they produced a pamphlet titled "The Poisons We Eat." "More and more people are aware of the poisons in the air, water, and food," the pamphlet read, and "boycotting grapes is one of the few ways the American consumer can act with proven effectiveness AGAINST pollution."[11] The federal law giving workers the right to organize excluded agricultural laborers, and many were undocumented immigrants who faced the threat of deportation, so allying with the public was especially necessary. In the end, by marching, striking, picketing, and boycotting, farm workers throughout Southern California (and elsewhere) gained union recognition and collective bargaining rights and negotiated contracts, giving them protection from chemical exposure well beyond any state or federal law at the time. There were criteria for field re-entry periods, protective safety gear, sanitation facilities, and medical monitoring, and various provisions prohibited the use of DDT, aldrin, dieldrin, and endrin, and restricted the use of other pesticides, a benefit, the union pointed out, "to workers and consumers alike."[12]

Awareness of the ways economic exploitation, racial prejudice, and environmental injustice were interlayered was also evident in campaigns by African Americans living in the country's decaying urban centers. In the middle of the 1960s, for instance, Black St. Louis residents drew on their considerable experience as civil rights activists to address the lead exposure that was peeling the paint off city "slum" apartments (often owned by White absentee landlords). They set up blood tests for thousands of children and pushed public officials to enact and enforce lead paint legislation there. Moreover, with assistance from White scientist Barry Commoner at nearby Washington University in 1969, they established the Environmental Field Program, directed by African American scientist Wilbur L. Thomas, Jr. In February of the following year, Thomas delivered a speech titled "Black Survival in our Polluted Cities" at Southern Illinois University, later published in *Proud* magazine for Earth Day, calling out the racist economic and political forces responsible for the environmental threats experienced by African Americans. "The nitty gritty issues relevant to Blacks is simply the fact that a disproportionate number of Blacks are exposed to more environmental hazards than non-Blacks, in addition to the regular burden," he explained, citing lead poisoning as a primary example.[13]

St. Louis passed its first lead law shortly after Earth Day, which proved inadequate to address the problem since the legislation split enforcement between the health department and the housing division. Organizers began direct-action protests, including sit-ins at realtors' offices and rent strikes against recalcitrant landlords. This eventually led to modifications in the law, although the changes still did not end lead poisoning, and the campaign continued and broadened at local, state, and national levels. In 1972, community leaders from across the United States convened a Black Political Convention in Gary, Indiana, hosted by Black Mayor Richard Hatcher, and their final platform included demands for "a living environment before profits" and an urban Environmental Protection Agency. Subsequently, in a report titled "Our Urban Environment and Our Most Endangered People," the EPA recognized this suggestion and acknowledged how the White, middle-class' shift to suburbs was sapping cities of resources and leaving those too poor to move to suffer the consequences.[14]

Racial exclusion was at the heart of another environmental battle, too, on Connecticut's "Gold Coast." There, wealthy White residents attempted to block Black city residents from gaining recreational access to beaches. Typically, the White homeowners worked through private "beach associations" and restricted access by limiting parking availability or banning street parking altogether, charging excessive fees to non-resident visitors, and erecting fences and seawalls (which disturbed local ecology). They also used zoning rules about lot size to prohibit multi-family housing and mobile homes, limiting who could afford to live in the coastal communities. Facing these obstacles to the shoreline's equal use, when an "Open Beaches" Act failed to pass in the US Congress in 1969, some Connecticut residents began direct action campaigns under the leadership of White organizer Edward T. "Ned" Coll. He had been focused on various urban problems throughout the state since the mid-decade and had an expansive view of their source, one more astutely mindful about how "White flight" was connected to so-called "urban pathology" than

McHarg's flawed layer-cake model of declining cities. "Violence in the streets," Coll declared in a speech, "starts with the yawn in the suburbs."[15]

One of the early protests was in Madison, where beaches were designated for town residents only, and Coll brought a group of city kids to stay in a local hotel. The town health director insisted on the children having a physical exam before visiting, however, and the Recreation Commission cited "insufficient information" to deny them guest permits. Subsequently, activists spent that summer and the following confronting residents at church services, town meetings, and the private beach club. Later, Black volunteer Anne Loiseau was attacked by two unknown assailants who slashed her face with a broken bottle. In Old Lyme, where activists had been attempting a similar access campaign, their local headquarters were vandalized and burned down. During the rest of the decade, as towns passed even more restrictive laws, civil liberties groups challenged them, arguing that using federal and state tax revenues to maintain beaches and other coastal features nixed barring non-resident access. Yet this, too, failed, as towns decided to forgo the public funding. "Let's be cautious that we are not trapped into using federal money in any way, shape or form to help pay off debt for purchases of our beaches or for their operations or maintenance," Madison select-woman Vera Dallas advised, "which would jeopardize restricting their use to residents only."[16]

A Matter of Definition

In the wake of the publication of Carson's celebrated book and McHarg's influential essays and book, many preservation, conservation, and environmental groups turned their attention away from working people and people of color and their concerns. Historians, journalists, filmmakers, and children's authors did something similar by occluding whole categories of people and groups from the standard narrative about the origins of American environmentalism. This happened, arguably, because of a narrow definition of who counts as an environmentalist and what counts as environmentalism, following the framing that *Silent Spring* and *Design with Nature* provided. Many of those who are part of the environmental movement or who are contributing to telling the environmental movement's story simply could or cannot imagine that workers, the poor, and people who are not White, might have an environmental sensibility or that they might participate or have participated in environmental protest. They also have trouble regarding the typically multifold concerns of working people, the impoverished, African Americans, Latinos, and others as properly environmental, discounting any effort that links environmental problems and economic or social injustice, or labeling it as something else. Consequently, activists are challenged to expand environmentalism's ranks and enlarge its agenda, and historians are challenged to see anything other than the flawed version of the environmental movement's founding and unfolding, one focused mainly on the White, middle-class and limited to their particular interests.[17]

1 Later, in 2008, Carson's biographer, Linda Lear, credited *Silent Spring* with giving rise to "a powerful social movement that would alter the course of American history." Marking the 50th anniversary of the book's release in 2012, the *New York Times* ran a story titled "How 'Silent Spring' Ignited the Environmental Movement." Furthermore, children's literature is suffused with hagiographic biographies, including titles like *Rachel Carson: Pioneer of Ecology, Rachel Carson: Founder of the Environmental Movement, and Rachel Carson and Her Book That Changed the World*. See Chad Montrie, *The Myth of Silent Spring: Rethinking the Origins of American Environmentalism* (University of California Press, 2018), 1–8.

2 Rachel Carson, *Silent Spring* (Houghton Mifflin, [1962] 1987), 30, 188.

3 Montrie, *The Myth of Silent Spring*, 99–101.

4 In 1964, McHarg published "The Place of Nature in the City of Man," expressing similar sentiments but with an urban planning perspective. "The most important inference from this body of information is that interdependence, not independence, characterizes natural systems," he wrote. "Thus, man-nature interdependence presumably holds true for urban man as for his rural contemporaries. We await the discovery of an appropriate physical and symbolic form for the urban man-nature relationship." See Frederick Steiner, "Preface," in Frederick Steiner (ed.), *The Essential Ian McHarg: Writings on Design and Nature* (Island Press, 2006), xii, xix–xx, and Ian McHarg, "Man and Environment" and "The Place of Nature in the City of Man" in Frederick Steiner (ed.), *The Essential Ian McHarg*, 1–2, 13, and 20.

5 Ian L. McHarg, *Design with Nature* (The American Museum of Natural History, 1969), 197.

6 Lewis Mumford, "Foreword," 20, and McHarg, *Design with Nature*, 193.

7 For some of the early scholarship along these lines see Andrew Hurley, *Environmental Inequalities: Class, Race, and Industrial Pollution in Gary, Indiana, 1945–1980* (University of North Carolina Press, 1995); Scott Dewey, "Working for the Environment: Organized Labor and the Origins of Environmentalism in the United States, 1948-1970," *Environmental History* 3, no. 1 (1998): 45–63; Robert Gordon, "'Shell No!': OCAW and the Labor-Environmental Alliance," *Environmental History* 3, no. 4 (1998): 460–87; and Chad Montrie, *To Save the Land and People: A History of Opposition to Surface Coal Mining in Appalachia* (University of North Carolina Press, 2003); Montrie, *The Myth of Silent Spring*, 8–16.

8 Montrie, *The Myth of Silent Spring*, 106–8.

9 Besides other support, the UAW paid for Environmental Action's nationally distributed newsletter. Additionally, the oversized banner on the main stage at the 22nd convention declared, "Enact–Environmental Bill of Rights... The Right to Fresh Air, Clean Water, and the Enjoyment of a Healthy, Attractive Living Environment." Montrie, *The Myth of Silent Spring*, 3–5, 108–9. See also Chad Montrie, *Making a Living: Work and Environment in the United States* (University of North Carolina Press, 2008), 91–112.

Beyond the need to get any interpretation of the past "right," the exposed fault lines in the environmentalism origin story matter because operating with a skewed historical consciousness impedes efforts in the present. Without a proper rendering of the past, we cannot confront the most pressing environmental problems of our time, like climate change and the need for a "just" energy transition. Failing to acknowledge the movement's deeper roots and push our historical perspective back to the start of industrial capitalism in the early 19th century, and to bring that perspective forward into the 20th century, we cannot see how labor exploitation, racial inequality, and environmental degradation were aspects of the same dramatic transformation. We also miss the pioneering role that countless ordinary people played in calling attention to these problems, often through unions, civil rights groups, and community campaigns. Only when we fully reckon with these social dimensions and a full roster of historical actors–in the United States and elsewhere–can we build a movement with a true capacity to address the challenges we face in the present and will face in the future.[18]

10 Environmental historians typically explain the origins of the environmental justice movement by reference to organizing at Love Canal in the second half of the 1970s and protests in Warren County, North Carolina, in the 1980s, followed by the First National People of Color Leadership Summit in Washington, DC, in 1991. Yet activists nationwide were already doing something like environmental justice by the early to mid-1960s. In addition to Mexican American farmworkers mentioned above, Southern Appalachian Mountain residents were fighting to ban surface coal mining that was ruining the landscape and impoverishing the region's economy. Their efforts, marked by an explicit link among environmental problems, economic inequality, and ethnic and racial marginalization, used grassroots community campaigns challenging corporate and government intransigence as well as militant tactics that included nonviolent civil disobedience and even industrial sabotage. Moreover, the UAW's meeting in 1976, "Working for Environmental and Economic Justice and Jobs," is likely where and when the actual phrase "environmental justice" was coined. See Chad Montrie, *A People's History of Environmentalism* (Continuum Press, 2011) and *To Save the Land and People*.

11 "Pesticides: The Poisons We Eat," Folder 20, "Pesticides, Leaflets, n.d.," Box 21, United Farm Workers: Administration Files Collection, Archives of Labor and Urban Affairs, Wayne State University, Detroit, Michigan.

12 Montrie, *The Myth of Silent Spring*, 16–18, 102–3. See also Montrie, *Making a Living*, 113–28, and Laura Pulido, *Environmentalism and Economic Justice: Two Chicano Struggles in the Southwest* (University of Arizona Press, 1996).

13 Montrie, *The Myth of Silent Spring*, 21–22, 114–8.

14 Ibid.; See also Robert R. Gioielli, *Environmental Activism and the Urban Crisis: Baltimore, St. Louis, Chicago* (Temple University Press, 2014), 38–39, 54–55, 142.

15 Montrie, *The Myth of Silent Spring*, 122–27. See also Andrew W. Kahrl, "Fear of an Open Beach: Public Rights and Private Interests in 1970s Coastal Connecticut," *Journal of American History* 102, no. 2 (2015): 433–62, and *Free the Beaches: The Story of Ned Coll and the Battle for America's Exclusive Shoreline* (Yale University Press, 2018).

16 Montrie, *The Myth of Silent Spring*, 122–27.

17 Ibid., 19.

18 For a consideration beyond the United States, into Canada, see Chad Montrie, "'What is Labour's Stake?': Workers and the History of Environmentalism in Alberta," *Labour/Le Travail* 93 (Spring 2024): 25–53, and "'Agenda for the 1970s': A Genealogy of Organized Labour's Environmental Activism in Ontario," *Papers in Canadian History and Environment* (September 2024).

IN CONVERSATION WITH JESS CONARD

Jess Conard is a longtime resident of East Palestine, Ohio, who was driven to grassroots advocacy following a recent train derailment and chemical accident. Her work and community advocacy champion medical attention, air monitoring, and implementation of local, state, and federal policies to clean up the toxic aftermath following the devastating damage from the spill of vinyl chloride (a chemical used in plastics production). As Appalachia director at Beyond Plastics, her new role pushes this work into national and international venues to push and advocate against plastics pollution and false solutions, often commonplace in industries that profit from having more plastics in the environment. **Jessica Varner** spoke with Conard about her personal experience of the derailment, the induced explosion that followed, the connections between vinyl chloride and the plastics we all use, the Midwestern erosion of truth in Ohio, the heartbreaks in community advocacy, and the lessons we can all learn from a resident-turned advocate in the face of growing global plastics pollution epidemic.

+ As someone who is also a fifth-generation rural Nebraska resident, trusting and listening to local, multi-generational voices is crucial to knowing about rural places and about rural people. In a country in which focus and expertise often fall on the coasts, the middle (or Midwest) continually finds folks muted or omitted from conversations. It is vital that folks get involved and advocate for the communities they are from, like you. You did not begin as an environmental activist or public health advocate per se. Your role as a community advocate is new following the East Palestine, Ohio train derailment on February 3, 2023. Can you introduce yourself and what inspired you to get involved in a complex situation?

I am the Appalachia director for Beyond Plastics. Prior to that, I was a licensed medical speech-language pathologist. I am essentially a lifelong resident of East Palestine. My family is here, and I attended elementary through high school here. I went away for a few years for college and decided to come back and raise my family here. Many generations came before me. I live on a road that is named after my ancestors. My family is very active in youth sports, community cleanup days, garden clubs, and all sorts of different community activities, but I was not always an environmental advocate. I have always been very active as a volunteer within the community, but not necessarily in the political community. I feel like before the derailment, I was living in a bubble. I was unaware of many government roles and how specific agencies are charged to protect us. I just assumed they would be there when you needed them. You hear about emergencies like pipeline explosions, the Flint water crisis, or the BP oil spill, and you think, oh, well, government agencies are there for them. They're cleaning it up; the experts and agencies are taking care of it. Our situation here in East Palestine received national attention, which a lot of disasters do not. I am grateful that we had a national platform because I truly think that East Palestine can be a catalyst for change: for emergency response, medical services, the way that we assess air, water, and soil pollution, and the list goes on. In the aftermath of the derailment, I learned that great policies and legislation already exist to manage these situations, yet we don't implement them because they cost so much money.

I also discovered that you can draw a straight line from what happened in my community to the plastics industry. When I figured out those connections, it gave me motivation and an understanding that I wasn't alone. Other communities had been handling similar disasters and emergencies out of the public eye. At that moment, I took it upon myself to reach out to these communities and ask what strategies they had implemented in their communities when the incidents happened or when there were releases of chemicals. In the process, I discovered an incredible, supportive network of people, and I was able to support them in return.

As a movement, we will not continue to allow the government to subsidize the plastics industry. This is the fight, right? The government doesn't want to fund the emergency measures needed to clean up the East Palestine disaster, and yet we continue to subsidize the production of plastics. It's a backward systemic issue, and a movement is underway to change the system. So, I proudly jumped into the national campaign to ban vinyl chloride. Beyond Plastics recruited me to help lead this campaign, first as a volunteer and then as the Appalachian director.

+ Can you describe what your community, landscape, and the environment are like? People, internationally or not familiar with the unique context–physically and socially–of small towns in the US, might not have a reference for what East Palestine, Ohio, is like.

East Palestine is a lovely rural community with about 4,700 people. We have a school, a grocery store, a post office, and a library. We don't have a hospital. Everybody knows everybody's business. Before the derailment, I had never worried about strangers coming into the town. The town is off the beaten path. We are very kid-focused, with many youth sports and opportunities for children. It's a sleepy town; we don't have a whole lot of change. Originally, East Palestine was an old pottery town with an industrial past. Most of the factories left when they rerouted the highway. But the train is a part of the culture here. The whistle is in every memory I can think of, every park family reunion, every backyard party barbecue. Everybody lives near the train. It goes right through our town. I never thought twice about what was on it.

+ I have seen a photograph in your backyard showing how the railroad runs directly behind your property. Can you describe the day of the derailment? Where were you? And what was most surprising to you about the incident?

The train tracks run about a football field from our backyard deck. The train was on fire when it passed by my home on Friday, February 3rd. The official report was that the train was on fire as far as Salem, Ohio, four towns away from East Palestine. There were sightings of the train as far back as Sebring, Ohio, which by car is about 45 minutes away. The first 911 call came in from Sebring, Ohio, so there was ample time to warn or stop the train before it made it to East Palestine, but that did not happen. Instead, a blazing train went through my backyard on that Friday night during the evening. My curtains were closed, so I didn't see it.

Starting around eight or nine o'clock on Friday night, I heard sirens from emergency vehicles going up and down the road. I went downstairs to check what was going on, and my partner said that a train had derailed and that there was a really big fire. It's not unusual for a train derailment to happen. We turned on the television and checked social media. Based on the photographs, we knew the train was ablaze, but we weren't sure what was on the train at that point. There was a lot of talk and misinformation about it being malt liquor or non-alcoholic beer. In fact, there was a single burnt train car carrying Modelo beer, which was kind of funny now, thinking back.

The next morning, we heard more about the contents of the train. That is when I first heard about the vinyl chloride car. They said, "It's leaking, we're monitoring the situation." At that point, they only allowed cameras in certain areas for the safety of the first responders. They weren't sure what was in the tanks. We didn't understand why they didn't have the manifest.

I remember they announced Sunday that there would be an imminent explosion, a spontaneous explosion of the vinyl chloride car. We were told that they could not control it; there wasn't anything they could do. The messaging was very confusing; we weren't sure if it was a one-mile radius or, more specifically, a one-by-two-mile radius. There were photos on the news of it being like a circle, a square, or a rectangle. We didn't know for sure. There wasn't a good system to notify people other than knocking on doors. In preparation for the detonation, they evacuated a one-mile radius around the derailment (our home was within a two-mile radius). Because we were within the two-mile radius, my family chose to shelter in place, which is the guidance we were given to keep people off the roads. Finally, they called in the National Guard.

School was canceled for the week, and I work remotely, so I was home with the kids that Monday. I remember the warning that they would do the vent and burn at 3:30 p.m. I tried to wait it out. But at about three o'clock, I panicked. I told the kids, "Grab your clothes, grab your toothbrush, and put it in this bag. We are leaving." I frantically threw whatever I could from my dresser drawers into a bag–family photos, my computer, and a book that I really liked. I remember thinking, this is all we are going to have left because my house is going to explode. I kept imagining the worst–the train car exploding like Chernobyl in my town and turned to ash. My fear came from a failure of public outreach and the ability to inform people about what was happening. I just wasn't prepared.

I have never driven so fast. I drove in the opposite direction toward Columbiana, Ohio. On the way out, I called my brother and demanded that he also get out. I panicked; I don't trust this. He was within a mile radius of the promised explosion. I was relieved when he said, "We're already on the road." We decided to meet at the Pizza Hut. All the cousins met around 3:30; the kids were not hungry and didn't want pizza. I said, "We're ordering pizza," and told everyone to sit down. We had all this food, but no one was eating. My brother was a local council person, so he was getting more updates than I was. There was a moment when he said, "They're calling it a success; I think we can go home now." I wondered, what does that mean, "success"? They had drained five of the cars containing vinyl chloride and set them on fire. I was stunned; why did they drain five? I thought it was just one leaking car. Reluctantly, we drove home. I remember seeing this apocalyptic plume when we arrived. It looked like a tornado that had just emerged out of the ground. The smoke plume was so widespread, and the smell was horrific. It was a terrible chemical burn smell–a scent that sends warning signals to your brain: get out, get out, get out. I took a couple of pictures. Knowing what I know now, I wish I had taken so many more. All I can remember was that terrible, terrible smell.

+ Do you remember what state and local agencies told you about the smell? Was there adequate information?

The EPA told us the smell was not toxic and that we were okay. And that we just had to shelter in place and keep our doors and windows shut. We were told not to go outside (we have pets, so we had to). When we let our pets out that day, sparingly, it still smelled really awful. The smell lasted the whole week. On Tuesday, they gave the all-clear and lifted the evacuation. And a few minutes later, a train rolled through. It was shocking; the burning had just stopped. How is it even possible that the tracks continued to operate so close to the accident? It was traumatic to see another train. And then I realized they hadn't cleaned it up. The train rolled through on top of what we now know is 116,000 gallons of vinyl chloride, which created 1.1 million pounds of air pollution. And no cleanup happened for the next 30 days.

In the month following that, I became a high-profile advocate. I was on CNN with Jake Tapper. I spoke to Erin Brockovich. I started speaking out, in interview after interview after interview. I went to New York. I went to Washington, DC. I sat in the Senate hearing behind Alan Shaw (CEO of Norfolk Southern Corp) as he apologized, saying, "I'm so sorry." Sorry. Sorry is what you say when you bump into someone. Sorry wasn't enough. He didn't look back at us. He didn't try to meet us. He didn't try to shake our hand. He didn't make eye contact.

+ I appreciate you sharing your detailed, emotionally charged account. Moreover, I'm sorry you have to relive these things over and over again to demand change and cleanup from Norfolk Southern.

Thank you; I appreciate that. The most mind-boggling thing is that I do not think they would have ever cleaned it up if we hadn't complained. Think of how many other smaller disaster residues remain after over the 150-plus years the rail has operated in our region, the 60 to 70 years of the plastics industry's polluting ways, and the same with oil and gas development. The structural issues ingrained go on.

+ What has been the most illuminating for you since the emergency in 2023? What shocked you most about learning more about the system of transporting hazardous materials in the US, about the governance of plastics, and about the role of federal governance and corporations?

I learned that the current system doesn't work for human health, human safety, and the environment. We have clearly and transparently demanded what we need from our government. Those pleas after 22 months of asking for very specific needs are an organized medical response, air monitoring, soil testing, well water testing, well filters, and drinkable water. In response, we received a $25 million park. Norfolk Southern also created a $16 million safety building in a renovated train station. Our teachers received $50,000 in gift cards. Our football player helmets had train tracks put on them. Norfolk Southern has held festivals. Yet, the basic human health requests go unanswered. I am still paying for my son's new asthma diagnosis and inhaler, which is $150 a month. I paid $5,000 for my home well water filter but was denied reimbursement from Norfolk Southern.

I was one of the advocates who wrote the letter inviting the president to come to our community. He came in February 2024. I met with President Biden. I read the letter I wrote and he listened. And yet, we still don't have what we need. We still don't have an emergency declaration that would create action policies that our tax dollars charged legislators to write, put into law, and thoughtfully enforce. We can't take advantage of it because we haven't been granted an emergency declaration. In other words, there are policies in place to govern plastics federally, and safeguards are in place to start the required due diligence to make the systems safer, but they're not being utilized as they should. Communities aren't getting what they asked for and what they need.

+ This incident seems to conflict between Midwestern values emphasizing belief and trust and the on-the-ground actions of large government and corporate entities. How has trust changed in your community in light of what happened? And where do you see the erosion of trust manifesting itself in East Palestine?

We are a look-you-in-the-eye, handshake community. If you say you will do something, I trust you will do it. In the context of this derailment, we have learned that there is a language game in how the government and the government entities present information. Even language used in simple testing reporting. They say, "There are no chemicals of concern, or it's below the reportable level." These terms are very misleading, especially when there is no safe exposure to the spilled chemicals. There is no safe level of exposure to known carcinogens like vinyl chloride and benzene. None of the chemicals on the train should be floating around in communities that don't have PPE. The chemicals should not be in our streams. They should not be in our soil. They should not be in our air. When you hear there is no level of concern, you assume your dog can drink the water. But that is not true because these standards are set based on a healthy white male. They don't consider pets or vulnerable populations. And they have no idea where these chemicals ended up. I could have a pocket of vinyl chloride in my backyard because they're not doing any soil testing. My children play in my yard, and they could fall into a puddle of vinyl chloride. We wouldn't know because we haven't had the opportunity to assess the impact's magnitude.

We had a private researcher test our pond water and our well and found evidence of vinyl chloride. The testing also found vinyl chloride on our back porch in a canister test. We brought it to the EPA, and their response was, "Oh, that's below the reportable level." I don't want my children hanging out on my back porch when there's vinyl chloride out there. Do you? They were also doing soil remediation at the derailment site without notice. They were disturbing the soil. I developed a rash, and they claimed no work was being done. Okay, I thought, what is soil remediation then? It's about language games. I think Midwesterners are not used to asking probing questions. We take it as it's given; I believe and trust what you say. Politics have changed here. The incident and its aftermath have interrupted families–more divorce and family separations. People ask, do we move, do we not move?

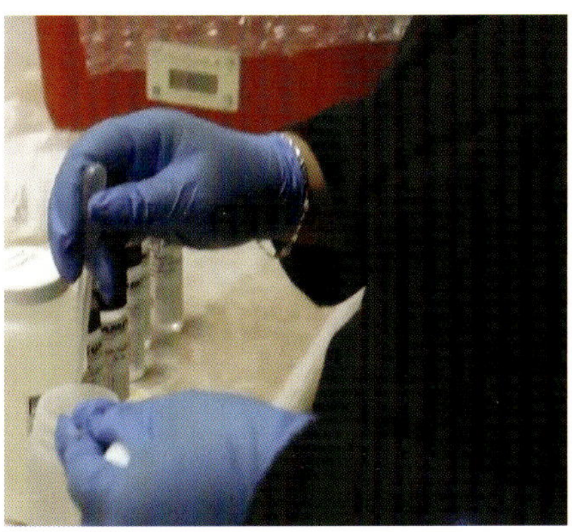

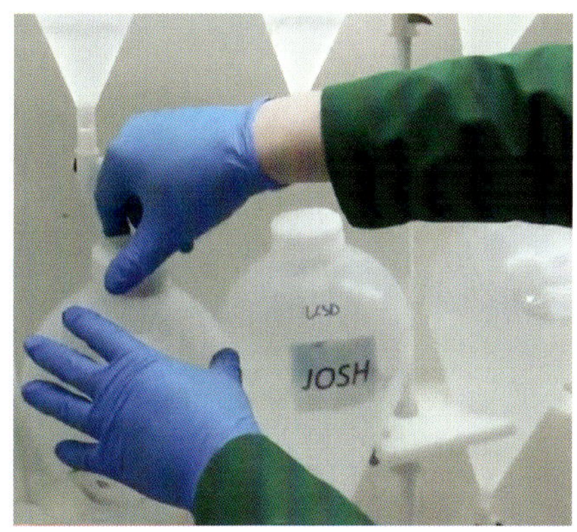
JOSH

+ Has there been a lot of movement from the town since the derailment?

I've lived here for a long time; it is a special place. People don't move. I have never seen this many houses up for sale. There are multiple buildings downtown that are for sale, as well as empty buildings. Many businesses have left (the last time I counted, 16 have left). And we don't have many businesses here in our small community. My mom has a small business, and she was forced to relocate. Some people have just cut their losses.

+ I read about President Biden's lack of response and how conservative commentators really used that moment to sow doubt. Can you describe your town politically before and after the derailment? Has the changing political landscape changed your political voice?

East Palestine has always been a ruby-red community. I grew up with Republican parents and adopted Republican values as my own as an adult, especially when I moved back. My values have shifted to align with more Democratic issues, but I also have a huge issue with the two-party system. Not everything is binary; politics are not black and white. We should have the opportunity to vote for the right person and not a party. That stance has gotten stronger since the derailment. The environment does not have a political party.

Biden was not here in person immediately following the accident. However, there was an active federal response right away. Research grants were handed out, but we shouldn't have to be guinea pigs. We have policies; we just need to implement them. For example, we have a rail safety bill that can't even make it to the Senate floor for a vote. What is the point of voting someone in if they are not going to work to put policies through to the finish line? We went to Washington, DC. We went to the State House. We went to New York. We met politicians in every place, and when they came here. We have communicated our needs. They have a responsibility. That is their job. It is infuriating when politicians are meek in pursuing what is right for the people. And why? Because of campaign funding. Our Governor, DeWine, gets a ton of money from Norfolk Southern. Decisions are based on that funding.

+ You have asked for actionable steps, such as less conflict of interest in testing, a mobile unit to help those with mobility and time-restricted schedules, and more. What is the progress? Where are you all now? Are you discouraged or hopeful most days?

A health clinic was paid for by tax dollars for one year (implemented by Governor DeWine). We were hopeful, but it has been a triage. The center does not specialize in identifying chemical health issues. It is not free. It is not open to Pennsylvania residents (the incident affected Ohio and Pennsylvania residents). The first week it opened, we inquired: Are you doing vinyl chloride testing? Are you looking at asthma? Can I get my treatments here? No was the answer to everything. We still have rashes. We still have respiratory illnesses. We have other things that are popping up. And on a personal note, I am getting ready to get tested for lymphoma. I'm 39 years old. I've never had this issue. I exercise, I eat well. Whether or not it is related to the derailment, the CDC did not give me any hope when it came in and held a community meeting and said, "You have been exposed to the chemicals, but don't worry, your cancers are treatable."

So, what is progress? We have a new park. There is a big building downtown that has the Norfolk Southern logo on it. We have a fun festival that we can go to every year. As an advocate, it is very ostracizing because I am told I'm not grateful. I've lost many friends from speaking out. But here I am in my community, and I keep fighting. Getting that emergency declaration pulled across the line would be progress. We could start pushing key programs through, people could relocate, or they could have medical issues covered. It is so simple, and it is just sitting on the President's desk.

We are changing the course of history. There is a reason why they did the cleanup after 30 days. It's because we went to Washington, DC. There is a reason the emergency declaration is on the President's desk. Because we have made actionable, incremental steps along the way. The National Transportation Safety Board has supported us from day one. There are people who are doing good work in government, and it's important to continue to see that. That is why I keep voting.

+ What are some ways that everyone can begin to advocate, in the space of oil and gas, in the larger supply chain in plastics, in the transportation of hazardous goods, and in the structural problems along the way? Are there ways that everyone can get involved?

Yes. Every global change starts at the grassroots level. Understand your community and create something that makes sense for it, whether it be a community garden or a trash cleanup. It's not something that you have to do out of pocket. It's hosting a natural Christmas decorations gathering at your library to avoid plastic tinsel and other plastic goods. You can write a letter to the editor, or call your local elected leader, or present something at your local council. The list goes on and on.

+ You are now part of a larger nonprofit, Beyond Plastics, which likely drives your focus on advocacy beyond the community level. How has your voice within a larger organization shifted regarding scale, and what policies or environmental initiatives do you aim to shift?

Beyond Plastics is an organization that seeks to educate–whether that be media, policymakers, or the public–on the plastic pollution crisis. As the Appalachia director, I work with grassroots groups to oppose false solutions to plastic waste. In addition to leading the national campaign against vinyl chloride, I also work to build coalitions in Ohio, West Virginia, and Pennsylvania. Recently, I started working in the Gulf South.

One way Beyond Plastics has succeeded is by working with legislators and policymakers at the local level to pass moratoriums or bans to help curb plastic production. For example, in Youngstown, Ohio, we worked alongside folks fighting a proposed pyrolysis plant where they pyrolyze tires, plastics, and e-waste; the city council passed the first-ever moratorium in the nation. The same strategy was implemented in Point Township, Pennsylvania, for the largest chemical recycling proposal in the nation, which also backed out after we started passing legislation in surrounding communities.

We work with policymakers, but it takes that local push first to educate about the truths of solutions to plastic waste and dampen false solutions that perpetuate the production of plastic, because we know that less than 10% of plastics are being recycled and that the idea of recycling (even if it is false) leads instead to producing more plastics. More specifically, I help to ensure that false plastics solutions aren't funded by the government. We are looking at this issue in Erie, Pennsylvania, with the Department of Energy. And to step back, when I say that plastics are subsidized, I mean our taxpayer dollars go into perpetuating plastic production. If our taxpayer dollars instead subsidized hemp or soy or mycelium, think about the equitable transition that would occur. Right now, more natural or bio-based products are expensive, right? There is a just and equitable transition to a more sustainable way of life–the so-called green tax. But we have the capacity to do better, and we must work to incentivize materials that replace plastics.

IN CONVERSATION WITH

MARIA IVANOVA

Maria Ivanova is the director of the School of Public Policy and Urban Affairs and professor of public policy at Northeastern University. Her work focuses on the science–policy interface, particularly related to international environmental institutions and agreements. Ivanova served on the first-ever UN Secretary-General's Scientific Advisory Board under Ban Ki-moon (2014–16). She has also engaged in the negotiation process for a new global treaty on plastic pollution as a delegate of Rwanda, the International Science Council, and now as head of delegation for Northeastern University. **Karen M'Closkey** spoke with Ivanova about her recent book, *The Untold Story of the World's Leading Environmental Institution: UNEP at Fifty (2021)*, and the necessity and challenges of global environmental governance.

+ You've spent months in United Nations (UN) archives and interviewed over 200 current and former UNEP staff members and countless others, including current and former UN secretaries-general. What prompted you to tell the story of UNEP?

The United Nations Environment Programme (UNEP) is what I call the anchor institution for the global environment. Two key achievements of the 1972 UN Conference on the Human Environment, which took place in Stockholm, were putting the environment on the global agenda and creating the institutional structure to deliver on that–UNEP.

Despite its pivotal role in global environmental governance, UNEP has remained surprisingly understudied compared to institutions like the World Bank or the IMF. My goal with the book was to fill this gap, providing a comprehensive guide to what UNEP does and how it operates. It is essentially a handbook on what UNEP is as an institution and how governments can work with it. I decided to write the book so that people could understand it and engage with it better. I have had the honor to work with several governments and deliver lectures about UNEP to governments represented at UNEP meetings, whether in Nairobi, Geneva, New York, Washington, DC, or the other regional offices. Political leaders and ambassadors have said that they need this book. Many of them are from ministries of foreign affairs, but even those from ministries of environment benefit from knowing more about the institutional architecture of UNEP. They need to understand the dynamics to advance the interests of their own country and, even more importantly, the global, collective interests of citizens on this planet.

+ How does UNEP work? Is it a mediator among various parties? Does it set agendas for other UN work or set legally binding agreements?

UNEP operates like any institution whose mission is to deliver on the interests of a collective. Just as the 1970s saw the creation of national bodies like the US Environmental Protection Agency (EPA) and a whole set of laws, including the Clean Air Act, the Clean Water Act, and the Endangered Species Act, UNEP was established to address environmental challenges that cross national borders. The United States assumed the leadership position of creating this entity at a global level. The vision for a global body on the environment is to identify the problems and reasons behind these problems and to report on the state of the global environment continually. In many ways, UNEP is meant to be the canary in the coal mine. Scientists identify a problem, and then UNEP acts as a convener, bringing together representatives across other UN bodies–whether agriculture, industry, or trade–to find ways to address that environmental issue and to bring these various interests together into collective action for global good on the environment. UNEP identifies the different policy options, and governments can implement them as they see fit. That is why I call UNEP the anchor institution, not the umbrella. UNEP does not tell any government what to do–it is not an enforcer–but the north star's goal is to ensure that the health of people and the planet is protected.

In essence, UNEP operates as a facilitator, convening governments and UN bodies to address environmental challenges collectively without imposing mandates. It helps to align diverse global interests to advance sustainable solutions, making it the anchor that guides collective action without dictating terms.

+ Can you explain the relationship between UNEP and the UN Environmental Assembly (UNEA)?

UNEA is UNEP's primary decision-making body. It convenes representatives from all member states to align their national interests with the global environmental agenda. UNEP started with a smaller governing body–the 57-member Governing Council–but after reform efforts, UNEA was established as a universal governing body with all 193 UN member states sitting at the table. Governments agreed to make the change in 2012 at the Rio+20 conference in Brazil, and the first UN Environment Assembly took place in June 2014.

+ What was an early and significant international effort to define an environmental issue as a global concern?

One such success is the Ramsar Convention on Wetlands, which exemplifies how global cooperation can yield tangible environmental outcomes. The convention was signed in 1971 in Ramsar, Iran, before UNEP was created. It is one of the most significant international agreements for environmental protection, but it has yet to become well known. The convention pertains to wetlands of international significance, and a whole set of regulations protects them.

I study various international environmental agreements, and my research team and I have developed the Environmental Conventions Index, which measures the extent to which countries implement the global agreements they sign. The Ramsar Convention is one of the most positive cases. Though it has the largest number of questions that countries must answer regarding whether and how they are implementing its measures, it also has the highest reporting rate. And many countries are fulfilling their commitments.

+ Why has it been so successful?

It has been successful for three reasons, one individual and two institutional. On the individual side, people–scientists and other networks created by people who care about wetlands–were vocal about what happens when they are destroyed. People fought to protect and build with care around them through simple things like boardwalks on stilts rather than paving. On the institutional side, the Ramsar Convention's secretariat helps countries to report. It collaborates with them, entices them, and assists them to the best extent possible to report and to do so regularly. It is critical to have that institutional support. That support has resulted in the third reason: a global network of scientists and regulators is now working on wetlands. And hopefully, landscape architects, too!

Ultimately, the Ramsar Convention Secretariat's proactive collaboration with countries has been a cornerstone of its success. It provides crucial assistance in reporting and fosters a global network of scientists and regulators committed to wetland conservation.

+ The Kyoto Protocol (adopted in 1997) was the first international treaty to commit nations to reduce greenhouse gas (GHG) emissions, and in 2015, the Paris Agreement was adopted. Both agreements come under the United Nations Framework Convention on Climate Change (UNFCCC). What are the relevant differences between the two treaties? Has one been more successful than the other?

Many people often say that work on climate began with the Kyoto Protocol, but that is inaccurate. The work started with the UNFCCC, one of the conventions adopted at the Rio Earth Summit in 1992. The Kyoto Protocol is one of the agreements that makes the convention more operational. The difference between the Kyoto Protocol and the Paris Agreement is that the Kyoto Protocol splits the world into two–those that are industrialized, the so-called "developed countries" that would be subject to GHG emissions mandates, and those that are "developing" and would not have those requirements. This is why the United States refused to sign; China's high emissions were considered "developing" and thus not subject to any requirements. Yes, its emissions are high, but we must also ask why. China is the manufacturing space for most of the world. Even though emissions originate in one country, it does not mean that the manufactured products are used in the country of production. So, it was not particularly helpful to divide the world this way, which is how the Paris Agreement came to be. All countries agreed to do something to reduce their impact on the climate, but it was up to the individual countries as to how they would do it.

Under the leadership of Christiana Figueres, the executive secretary of the UNFCCC from 2010–16, the narrative around climate action shifted from one of sacrifice to one of opportunity. She emphasized the link between climate action, economic growth, and social benefits, transforming the discourse and laying the groundwork for the Paris Agreement. Figueres linked the values and deliverables in fundamental ways. This allowed people to see the importance of doing what is right for everyone while accommodating different countries' energy production methods and paths to energy transition. Some people say this is the worst part about the Paris Agreement–that nations get to set their own goals and programs and have nationally determined contributions–but others argue that this is the best part of the agreement. The key issues are who is monitoring, who is doing what, and what we are learning in the process. That is where I see that academia has an important role, which we have not entirely embraced yet.

+ 1992 was a big year. The Rio Earth Summit saw the creation of the UNFCCC and the Convention on Biological Diversity. Climate change and biodiversity loss are closely linked and mutually reinforcing. Has liaising between the two been successful such that programs to address one are not detrimental to the other, such as Reducing Emissions from Deforestation and Forest Degradation in Developing Countries [REDD+], which has been criticized for encouraging plantation forestry for carbon offsets? This condition is devasting for biodiversity.

This is a challenge at both national and global levels. At the national level, there are various ministries–Ministry of Environment, Trade, or Food and Agriculture–and there are very few instances where they all come together and see the nexus issues and the trade-offs. What UNEP has done most recently under Inger Andersen, the current executive director, is to home in on the narrative of the triple planetary crisis: climate change, nature loss, and pollution. They cannot be addressed separately. So now UNEP and many of the institutions and programs of the United Nations have taken on this narrative as an ambition, as the way forward. A lot of discussion is happening among the different environmental conventions, as well as between environmental conventions and, for example, trade conventions. Regarding biodiversity, there is the CITES treaty [Convention on International Trade in Endangered Species]. On pollution and chemicals, there is the Basel Convention on the Trade of Hazardous Waste or the Stockholm Convention on Persistent Organic Pollutants. How can action on these challenges be related? I am seeing more and more deliberate discussions and actions around that question. The Summit of the Future in New York [September 2024] was a step in this direction. The key outcome document was the Pact for the Future, which brings these issues together.

+ The UN Sustainable Development Goals [SDG, 2015] are lauded by some but criticized by others for vagueness and exacerbating inequality. The latest Convention on Biological Diversity [COP15]–which aims to conserve 30% of land by 2030 and is colloquially known as 30x30–has four goals and 23 targets. Does this have a similar chance of looking good on paper but less so in practice, or do you not agree with that assessment of the SDGs?

I do not agree with that assessment because creating global goals is not an easy undertaking. I have had the honor and pleasure to learn about the creation of the SDGs from the two women who made them happen, Paula Caballero and Patti Londoño. They wrote a book about the SDGs, which tells a powerful story about how they came up with the idea and how difficult it was–as two women from Colombia–to get it through the UN. Now that we have the goals, it is easy to criticize and say what isn't working. But what are the options? SDGs were a tremendous endeavor–they broke all the traditional divisions of the UN in which countries belonged to certain blocs, like the G77, which are the "developing" countries. The SDGs were created by an open-ended working group and any country could be a part of it. This is perhaps why we ended up with seventeen goals, which may seem too many.

The seventeen goals are ambitious and complex, and we have not achieved them. The 30x30 is a similar construct of focusing our energies. Some say, "Why bother setting these goals if we cannot achieve them?" But we do not know where we are headed if we do not set any goals. So, I will repeat it: we need academia in this space. We need a much more thoughtful way of reviewing and analyzing how goals were set, what has resulted from creating these goals, and how we might do things differently. We could also compare across various settings, for example, reviewing the Aichi biodiversity targets or 30x30 through the SDGs. But I think we need much more serious academic analytical work to identify what has worked and has not. However, we cannot have global governance without goals.

+ The UN has 193 member states, ranging from more democratic to more authoritarian, and Indigenous people are not recognized as sovereign under UN membership. Can we have progress on global environmental issues without genuinely democratic forms of representation, especially in terms of how these goals impact local people and landscapes?

How do we define democracy, or say this country is a democracy, and this one is not? I prefer to think about it in terms of good governance rather than whether a country is defined as a democracy. Is having elections every two- or four-years good governance? Putting the collective's interest above the individual's interest and creating the necessary institutional, regulatory, and legal structures, systems, and processes is imperative. However, it is tough to harmonize regulations and actions at the local and global levels. What is challenging is that inequity is baked into the nature of capitalism. How can you have a system of equity and delivery for everyone without radically changing the political and economic system? That is a much bigger challenge. In the meantime, we still need progress on environmental issues.

I work a lot with Rwanda, and I have seen what a country that emerged from genocide has been able to deliver for its population in just a couple of decades. The country has transformed and become a leader in environmental issues related to its population and global issues like plastics, which is my new area of professional focus.

+ Can you give a specific example in Rwanda?

There are many, including in wetlands protection, that we discussed earlier. Rwanda has improved its implementation of the Ramsar Convention, and the effects have been stark. Biodiversity has returned to the Rugezi wetland, the water table has lifted, and electricity is functioning because it runs on hydropower. The grey-crowned crane, Rwanda's national bird, has also returned as the ecosystem has recovered.

However, plastic pollution is perhaps the most important recent example of Rwanda's leadership. Two small states, Rwanda and Peru, co-led the resolution for a new global treaty to eliminate plastic pollution at the UN Environment Assembly in March 2022. The resolution was unanimously adopted, leading to the Intergovernmental Negotiating Committee process. I had the honor of serving on the national delegation of Rwanda, and we are now going through those negotiation processes. The name suggested for the global treaty on plastics is Kigalima, from Kigali and Lima, the capitals of the two countries that initiated this treaty. It will be the first UN convention that merges the names of two cities.

+ It seems surprising that it has taken so long to have a treaty on plastics, given how long we have known about plastic problems. Is it because of the latest science about the extent of microplastics?

The science on the extent and amount of plastics is one aspect, and the rise in certain cancers, especially among young people, is another. But I think the visuals of the plastic straw in the turtle's nose, the plastic on beaches, and in fish and birds also pushed plastics into the political sphere. Those visuals have been compelling and have forced people to say, "We can't go on like this; something must be done." Understanding the effect on human health is critical, but understanding is insufficient. We need to see it. We need to hear it. We also need to know that it is reversible. And that we can do something about it, individually and institutionally.

+ Kim Stanley Robinson's climate fiction novel *The Ministry for the Future* is set in the near future. The ministry has been formed under the auspices of the UN Paris Agreement. Its sole mission is to safeguard future generations by treating their rights as we would the rights of those already living. The book is a checklist of the many existing and hypothetical mechanisms for rapidly decarbonizing. This goal to protect future generations is inherent in many environmental initiatives, but is there a role for UNEP to be a think tank for actions that presently seem impossible or outlandish?

UNEP should be that. It was created to be that. UNEP was established as the "ecological conscience" of the world, and it has the potential to be much more. It should embrace a future-focused role. By partnering with universities, design schools, and think tanks, UNEP could help to foster innovative solutions to our current environmental challenges and those we have not yet imagined. Creating a futures lab would allow UNEP to explore new ideas beyond the constraints of current politics and to lead global efforts in designing a sustainable future.

I appreciate your question about *The Ministry for the Future*, because having science fiction, or just fiction, informs our thinking about possible responses to current and future challenges. And this is liberating. When we shift into fiction, we go beyond the constraints of the moment and ask what is possible rather than what cannot be done right now, which is what politics is. That blue sky scenario is empowering; I think this should be UNEP's role. Your question prompted me to believe that UNEP should have an active global futures lab and connect with universities and design schools. There is a UN Science Policy Business Forum based at UNEP in Nairobi, which brings scientists, government officials, and businesses together to innovate for a different world. This is something that should be done to design beyond plastics. There is a lot of space to innovate, but it needs to be curated and facilitated. UNEP would be a great organization to do that.

+ You begin your book with a quote from climate activist Greta Thunberg and describe how leaders have failed to progress on the environmental conditions for which UNEP was founded. In many cases, they have gotten much worse. Is there a recent development that has given you hope?

The move toward a plastics agreement has given me much hope. I have seen it shift from no agreement or a focus on marine plastics to a commitment to create a global treaty on plastic pollution that was ushered in by two small states (Rwanda and Peru) in just two years. And to see what that has done to several governments who have mobilized their industry, science, and civil society has been incredibly positive. Of course, there are challenges. Of course, there are hurdles. But just the fact that a process toward such an agreement has begun gives me hope, and seeing young people engage and re-engage in this work has been tremendously motivating.

Teaching environmental policy and governance has always been challenging because many young people say it is depressing and difficult. But I see a different attitude now: they're saying, "What can I do about it? How can I contribute?" Organizing through the children and youth major group at UNEP, for example, or through the Summit of the Future and a commitment by the secretary-general to create an agreement on future generations is akin to the declaration of human rights. On the one hand, we declare that the planet has rights and that we need to protect the environment. On the other hand, the future has rights, and we need to protect the environment for future generations to thrive.

MELISSA S. RAGAIN

ON THE ROAD WITH

HERBERT BAYER

Melissa S. Ragain is a professor at Montana State University, specializing in environmental aesthetics and modern art history. She authored *Domesticating the Invisible* (UC Press, 2021) and edited Jack Burnham's collected writings (MIT Press, 2015). Her work appears in major art publications, and she has received fellowships from Harvard's Radcliffe Institute and the Museum of Fine Arts Houston. Based in Livingston, Montana, she also curates special projects for Tinworks Art, focused on placing regional art within broader contexts.

+ ART HISTORY, DESIGN

The massive changes that environmental aesthetics underwent in the 20th century would have been unthinkable without two key concepts, drawn from biology and design, respectively: the ecosystem and the total environment. In the 1970s, the emerging science of "ecosystems" represented entities within a particular biome as part of a complex economy for which energy was the universal currency. This new paradigm expanded conservation beyond species and habitat preservation to include the analysis of the environment as a set of interconnected energetic systems, whose various component organisms needed to be understood collectively and metabolically in relation to surrounding material conditions.[1] The concept of a "total environment" was imported by European designers, such as the architect and pedagogue Eduard Sekler, who used it to describe a civic-scaled and human-centered approach to design that asserted complete control over forms and interactions within a given space. Unsurprisingly, Sekler's approach was pitched to future members of the political and technocratic elite at places like Harvard and MIT. Emerging from very different academic disciplines, the ecosystem and the total environment were nevertheless highly compatible, especially in the heyday of cybernetics, which tantalized the managerial class with promises of "command and control" over all manner of systems.

Significantly, the popularization of this kind of energetic model coincided with the growth of the worldwide energy economy. The United States and other developed countries relied heavily on carbon-based fuels like coal and petroleum, which necessitated elaborate systems of international trade and physical infrastructure to support the circulation and processing of valuable energy sources.[2] Ironically, petrocapitalists, hoping to secure global pathways for the redistribution of oil, relied on the same "whole Earth" rhetoric as their hippie counterparts. Further, this increasingly abstracted and holistic way of talking about environments fostered an understanding of the world as a single biome composed of fungible units of energy that could cross national and ecological boundaries. These projects exemplified the values of containment, circulation, and exchange that drive global economies and traverse many local ecological and geopolitical contexts due to their massive physical footprint. That also meant that by implementing market-based mechanisms such as today's carbon or cap-and-trade system, global actors could effectively delocalize the impacts of energy extraction and, at least conceptually, dilute real or potential harm by recontextualizing it within a much larger worldwide energy system.

In the decades that followed, the concept of "sustainable development" took hold under the watchful eye of a generation of (mostly) men connected to the University of Chicago and the Aspen Institute. These individuals, steeped in modern art and classical philosophy, believed that corporations could be the unlikely handmaidens of cultural unity, individual freedom, and ecological well-being. This "eco-establishment" expanded the urbanist's "total environment" using systems of global environmental management, thereby shaping environmental policies and practices.[3] The 1972 UN Stockholm Conference played a pivotal role in shaping environmental policy, codifying the idea that economic growth and capital accumulation were not just compatible with but were the necessary precursors of environmental regulation. This perspective, that the environment was a space to be managed, led to the belief that its management should fall to powerful Western nations and multinational corporations that also controlled massive infrastructural projects like pipelines, canals, and highways.

Art historians, including myself, have primarily focused on these changes within the milieu of experimental art practices of the 1960s and '70s. However, the aesthetic products of industry are equally relevant. Take, for example, the Bauhaus designer Herbert Bayer (1900–1985), who worked for his close friend Robert O. Anderson as the design consultant for Atlantic Richfield (ARCO). ARCO was the energy company responsible for constructing the 800-mile-long Trans Alaska Pipeline System (TAPS) from 1974 to 1977. Both self-described environmentalists, Anderson and Bayer developed now-common public relations strategies for the petroleum industry, chief among them their sponsorship of social welfare programs, environmental remediation projects, and the arts. Bayer's prolific work for ARCO from 1966 to 1985 cultivated its corporate image as a "dynamic, forward-looking energy company" while maintaining continuity with his existing practice of creating immersive aesthetic spaces. His ARCO work included the expected visual identity materials—logos, letterhead, and so on—but also painting, sculpture, textiles, a gallery, and a massive art collection.[4] I am interested here in Bayer's series of proposals for large-scale highway projects as exemplifying the converging aims of extractive enterprise and environmental design in the early days of neoliberalism. Made between 1969 and 1972, these proposals span the years between the Santa Barbara oil spill and the darkest days of the embargo, when American energy companies simultaneously confronted harsh criticism for pollution and fervent demands for free-flowing oil. Like ARCO itself, Bayer resolved these contradictory demands by using what I call "canalization" techniques that moved people or material through a space while minimizing interaction with the surrounding environment, thereby aestheticizing the channel or corridor as a paradigmatic modern form.

The view from the road

Bayer's first consideration of the highway came early in his American career, designing the dust jacket for architectural historian Siegfried Giedion's influential history *Space, Time, and Architecture* in 1941. The book is now considered a classic because it set aside questions of historical style in favor of an analysis of the spatial paradigms that informed modern architecture and because it integrated new discoveries in the sciences. Giedion appropriated the term "space-time" from theoretical physics to describe the kinesthetic experience of architecture since the Industrial Revolution. Forsaking the three-dimensional perspective of the Renaissance, he argued, modern architects and painters had embraced fourth-dimensional space-time, expressed through their use of simultaneity and movement. To illustrate this distinction, Bayer created a photocollage to represent these two spatio-temporalities on the book's cover. He used a red monochrome depiction of the formal gardens at Versailles, with its straight lines and central vanishing point, to epitomize the classical structuring of perspectival space. Modern space-time—what Bayer's colleague László Moholy-Nagy would call "vison-in-motion"—structured perception rhythmically by shuttling the viewer through a series of interpenetrating forms, which Bayer represented using the looping diagonal of Randall's Island cloverleaf superimposed in blue.[5]

The highway, not the masterpieces of International Modernism, represented the triumph of speed over solidity. The fourth dimensionality that characterized Giedion's "space-time" affected exhibition design as well as architecture. Famously, Bayer's experimental exhibitions of the 1930s comprised enveloping spaces that could engage the viewer's total field of vision by breaking away from the orthogonal surfaces of conventional buildings. As art historian Rebecca Uchill has noted, Bayer and his frequent collaborator, the curator Alexander Dorner, used the same theory of space-time to justify immersive installation practices that centered on the perceiving subject.[6] In his design for "Airways to Peace," a spherical exhibition space encased the static viewer within a 360-degree informational surround. Other immersive designs accommodated a mobile viewer. For example, the canted billboards in his 1936 model for an outdoor exhibition and the elevated walkways in "Road to Victory" (MoMA, 1942) controlled foot traffic by keeping viewers continually moving through media environments. If the exhibition attendee were an eyeball-with-feet, the driver was simply an eyeball-on-wheels, and the highway could be treated like a massively accelerated exhibition program.

This approach had been corroborated by researchers at the Joint Urban Studies Program at MIT and Harvard. Donald Appleyard, John Myer, and Kevin Lynch analyzed the highway system as a test case "to find a visual means for pulling together large urban areas."[7] Their findings, published as *The View from the Road* (1964), concluded that designers should treat the highway as a lesson in structured attention, "played to the eyes of a captive, somewhat fearful, but partially inattentive audience whose vision is filtered and directed forward."[8] Appleyard, et al., encouraged designers to consider how highways could produce orienting aesthetic experiences for commuters while also encouraging safe driving, and deliberately neglected the appearance of the highway from the "outside." As the highway pulled a viewer through edges, nodes, districts, and landmarks that made up the overall image of the city, designers could use the road to create meaningful experiences out of these elements. Design sequences might make use of sensations such as the motion of the self and the apparent motion of the surrounding environment, perceived compression or extension of the self, or the dramatic effects of approach and departure. Because the direction and grade of the road compelled the attention of the driver, they demonstrated, the driver's eyes were typically drawn to closer, apparently mobile objects as opposed to distant, seemingly stable ones. That meant that humanmade structures in proximity to the road made a far greater visual impression than the surrounding landscape, particularly objects that used simple shapes and flat surfaces easily seen by highspeed viewers. Above all, it was imperative that designers exploit "opportunities for visual emphasis" that would claim attention and rattle the driver out of a normalized state of distraction.[9]

Nothing to see

Inspired by Bayer's long car trips through the Great Basin on his way from Aspen to the ARCO headquarters in Los Angeles, many of his proposed highway sculptures purport to alleviate boredom and distraction by activating the driver's attention in otherwise flat, featureless stretches of desert road. In addition to canalizing and controlling vision, highways could recreate the kinesthetic pleasures of outdoor recreation. There are sweeping passages in *Space, Time and Architecture*, for example, that reveal Giedion's delight in roads that recall journeys through rolling foothills:

> Confidence is given to the driver in the way the road fits into the earth between its sloping sides. ...Yet he is held to a natural speed limit by the adaptation of the roadbed to the structure of the country, by its rise and fall, the smooth swing of its curves, its clear open runs before creeping under a cross highway or bridge. ...Riding up and down the long sweeping grades produces an exhilarating dual feeling, one of being connected with the soil and yet of hovering just above it, a feeling which is like nothing else so much as sliding swiftly on skis through untouched snow down the sides of high mountains.[10]

Both Giedion and Bayer were raised in Alpine ski cultures in Sweden and Austria, respectively, and both understood skiing as an activity that combined thrilling movement with restorative

1 See the author's previous publications on the impact of this shift on compositional studies: Melissa Ragain, "'Homeostasis is Not Enough': Order and Survival in Early Ecological Art," *Art Journal* 71, no. 3 (2012): 78–97; Melissa Ragain, *Domesticating the Invisible: Form & Environmental Anxiety in Postwar America* (University of California Press, 2021). Other historians of visual culture have also noted this shift; for example, James Nisbet, *Ecologies, Environments, and Energy Systems in Art of the 1960s and 1970s* (the MIT Press, 2014), and Douglas Kahn (ed.), *Energies in the Arts* (the MIT Press, 2019).

2 My thanks to colleagues at Petrocultures 2024 for introducing me to the prevalence of the word "abstraction" in the ways historians spoke about the carbon-based economy. I wish to acknowledge Derek Woods, in particular, since his presentation on the New Alchemy Institute helped me connect the abstractions of ecosystems science to delocalized treatments of the environment by energy companies.

3 "The eco-establishment" is a term coined by Katherine Barkley and Steve Weissman in Editors of *Ramparts* (eds), *Eco-catastrophe* (Harper & Row, 1970), 15–24.

4 This phrase is taken from Bayer's ARCO consultancy contract, Herbert Bayer Papers, Denver Public Library.

5 Hiller calls Moholy-Nagy's spiraling designs and kinetic art objects "therapy for democracy." However, it is unclear whether Bayer or Giedion shared this pedagogical intent. Instead, they treated space-time as a fact of modern life. Christian Hiller, "Vision in Motion –> Information Landscapes – From Stage Props and Camouflage Techniques to Democratic Apparatus and Cybernetic Networks," *Bauhaus Imaginista Journal* 4 (2019), https://www.bauhaus-imaginista.org/articles/4250/vision-in-motion-information-landscapes.

6 Rebecca Uchill, "Re-viewing *The Way Beyond "Art"*: Herbert Bayer, Alexander Dorner, and Practices of Viewership," *Architectural Theory Review* 23, no. 1 (2019): 114–45.

7 Donald Appleyard, Kevin Lynch & John R. Myer, *The View from the Road* (Published for the Joint Center for Urban Studies of the Massachusetts Institute of Technology and Harvard University by the MIT Press, 1964), 63.

8 American companies promoted a brand of landscape tourism that would have been familiar to Bayer from his Weimar days. Ford Motor Company, for example, hired the now well-known illustrator Charley Harper to design the covers for their consumer publication that highlighted the car as a way for individuals to access the natural world as a middle-class leisure activity, as opposed to the functional role played by public transportation. Harper's science-driven illustrations of America's many ecoregions helped solidify the association between a healthy environment and *Ford Time*'s vision of "life through the windshield." Appleyard, et al., *The View from the Road*, 5.

meditation. Many of Bayer's highway proposals provide the kind of exhilarating and restorative experiences described by Giedion above. Skiing, like driving, was a high-speed alternative to hiking, the preferred Bavarian method for communing with the natural world. Never moving in a straight line toward a determined end point, the skier descends gradually by a series of connected C-shapes, cutting an undulating line through space. The skier is never more aware of the terrain than when accommodating its ups and downs, an embodied experience that complemented the panoramic mountain view. Skiing and driving were not only modes of transportation; they were forms of recreation that allowed the sluggish human body to skim the broad, undulating surfaces of the earth. Ultimately, however, the natural world was a space to look at, move through or across, and to return home from, restored.

ARCO Refinery, Philadelphia, 1973 sculptural screens (proposal).

In Weimar Germany, where Bayer had spent his early career, automobiles were associated with bourgeois leisure activity and access to nature outside of the city limits. Nature, due to its associations with purity, cleanliness, and health, was a remedy for industrialized urban life. The mountainous landscapes of the Alps are reflected in Bayer's adamant protection of mountainous viewsheds in his adopted hometown of Aspen, Colorado. Further, Bayer understood environmental design as the process of protecting the natural world by maintaining a strict separation between humanmade and natural spaces, a separation the highway helped to enforce:

> We must learn to look upon nature with reverence as the source of all existence.... The artist will not imitate nature but create a spiritual world of itself, side by side with nature.... The structures which man erects will not compete with nature nor set themselves up against it. Both the natural environment and man-made environment will exist with each other if their boundaries are understood.[11]

Highway Gate in Three Parts, 1971 acrylic, collage, and graphite.

Even as leisure activities like skiing, hiking, or driving might move through natural spaces, design's task was to protect the purity of nature by creating a *synthetic* nature in the form of engaging, immersive abstractions. While developed with safety in mind, this category of highway proposal also responded to the perceived inadequacy of the landscape between the destination and the point of departure. In the absence of natural features recognizable to Bayer, synthetic variances in the landscape could serve the purpose of maintaining attention, heightening the sense of apparent motion, while providing embodied forms of entertainment. Monumental primary-colored gates to drive under, earthen mounds to weave through, and even terrifying gorges to traverse, all piqued drivers' attention using the elements of compression and extension, approach and departure, and visual and kinetic sensation that simulated the experience of moving through a

ARCO Refinery, Philadelphia, 1972 fence and building (proposal).

Road in deep embankments, 1972 acrylic and charcoal.

Highway landscape with four hills, 1972 pastel and collage.

Positive negative highway, 1971, wood, mat board, paint, toy automobile.

mountainous landscape. Bayer's highway structures provided a sanitized spatial logic that resisted enmeshment in the local cultural or ecological environment but did so in the spirit of environmental protection.

Something to hide

Many highway projects use prefabricated elements to produce what Bayer called "articulated structures." Bayer initially developed this visual vocabulary for his *Articulated Wall* on the *Ruta de la Amistad*, an ambitious series of highway sculptures commissioned for the 1968 Olympics in Mexico City.[12] The roadside context, inside a cloverleaf interchange, and material limitations had inspired Bayer to work in monochrome with simple repeated units that could create interesting effects of light and shadow for the passing driver. Using the accordion-like texture made by repeated rectilinear columns, he was able to create an organic form and dynamic effect that rewards movement around the object and takes advantage of the changing angle of natural light. Bayer created numerous variations on these articulated structures for ARCO, often drawing inspiration from ancient monumental architecture like cliff dwellings or pyramids. While some of his "symbolic" designs included the ARCO Spark logo, in others the repeated units obliquely reference the "techniques of transportability" so essential to the oil industry.

Red Pyramid and *Undulating Wall* belong to a second category of highway projects designed for a refinery in Philadelphia that ARCO had recently acquired in anticipation of approval to construct TAPS. Though the pipeline was delayed by several years due to lawsuits by Indigenous and environmental activists, the 1973 energy crisis led to a fast-tracked approval, sending vast quantities of crude to Philadelphia for refining. The refinery faced a nearby highway that connected the airport to downtown, creating a disorderly and dirty image of the company in the minds of the local public. Bayer proposed remaking the entire site. His official proposal to ARCO in 1972 to "aesthetically improve" the refinery included a new paint color-coded scheme for the tanks, rearranging buildings to make the site more legible to visitors, landscaping, and designing several monumental objects, such as *Red Pyramid*. While it reads as a single monumental object from afar, upon closer inspection *Red Pyramid* is clearly made up of hundreds of painted oil drums. These objects drew on German Mexican artist Matthias Goeritz's theory of "emotional architecture," which argued that architecture should affectively and spiritually exceed the requirements of pure functionalism preached by CIAM and the Bauhaus. Bayer called these spiritual-emotional forms "environmental" because they created atmosphere through light, form, and color, and incited positive and, at times, transcendent sensory experiences.

Bayer's redesign of the Philadelphia refinery attempted to create aesthetic cohesion and cleanliness of the site for its users. But the most impressive components of his proposal were directed at passersby. In the 1960s, Bayer publicly advocated the use of architectural surfaces such as fences and screens as important tools for revising and editing the industrial environment. He even cited the comic pastoral murals covering the exterior of Farmer John's, a pork processing plant on Vernon Avenue in Los Angeles, to demonstrate how a keen use of bare walls can create positive brand associations despite the unsettling reality within. Not surprisingly, Bayer turned to the screen to remediate ARCO's more unsightly facilities. In a rendering of one of his articulated structures, *Undulating Wall*, the bright yellow structure sits in front of the refinery, which Anderson had recently acquired for processing the Alaskan crude that would soon flow through TAPS. The photocollage shows Bayer repurposing a sculptural object as a screen to shield the plant from the view of passing traffic and suggests the monumental scale he imagined for this work. The towers of a typical refinery are around 30 m tall. He supplemented the monolithic structures with canted screens and fencing covered in colorful dazzle camouflage. He even suggested hiding the refinery from airplane passengers by covering the site with camouflage netting of the sort used to hide soldiers from enemy planes.

From the road, drivers could experience the changing light effects across these engaging abstract objects as well as brightly colored murals that functioned both as safety devices–they drew attention to one's actual speed and warned of upcoming changes in the terrain–and as diversions from the towering plant on the other side of the retaining wall. The effect, again, is of a short but immersive thrill ride, a synthetic environment built for human viewers that trains the driver's attention on the act of moving through rather than tarrying within a given environment. Collectively, the highway projects maximize the joy of moving from point A to point B while screening out the messy local realities of the extraction that powered that very movement.

Conclusion

In the years following the Santa Barbara oil spill, the public newly understood that oil was not only dangerous to communities near the extraction site but also along its corridor – in the pipeline, on the tanker, or at the refinery. It was in the long stretches that crossed the permafrost that most of the Indigenous resistance to TAPS took place. Nevertheless, that awareness depended largely on the public's access to images or first-hand experience of spills. "People," ARCO Chairman Robert O. Anderson wrote, "consider petroleum an environmental hazard for largely aesthetic reasons, even though petroleum is a natural constituent of the earth, one that nature's creatures lived with long before the modern age and its oil spills."[13] To remedy this aesthetic problem, ARCO needed to emphasize oil's containment while diverting the public's attention away from sites of contact with its messy materiality.

In her analysis of plastic and other petrochemical products, art historian Heather Davis has described synthetic materials as "created irrespective of their surrounding environment." This process of delocalization, she argues, alienates the consumer from the materiality that connects petrochemicals to their earthly sources.[14] Synthetic materials help make the violence of extraction acceptable by "forgetting or rendering invisible the infrastructures, ideologies, labor, and materials essential to their production."[15] Bayer's highway proposals, while made of conventional materials like concrete, conform to Davis's definition of the synthetic. They are also symptomatic of the canalizing of the environment, where the creation of large-scale managed environments prioritizes frictionless circulation over slower moving, more complex environments that embody the "mesh" or "web" of life. His works appealed to postwar powerbrokers obsessed with fungible aspects of ecosystems and the managerial promise of total environments by promoting a vision of a human-centered world made possible by vastly expedited movement and exchange. The expediency of the canalized environment almost always benefits the nonlocal to the detriment of the local. Drivers on the highway cannot see the blighted neighborhood they bisect; oil flows within a prophylactic pipeline across flyover country that is also someone's homeland. Moreover, as a paradigm for exhibitions or infrastructure, the canalized environment splits its audience in two: those for whom the corridor exists, whose views are protected by its deliberately diversionary architecture, and those who view it from outside, the unintended viewer whose sidelong glance threatens to puncture the constructed environment within.

9 Ibid., 6.

10 Gideon, *Space, Time and Architecture*, 553–54.

11 Herbert Bayer, "On Environment," *New Mexico Architecture* 4, no. 5 (1962), 13, https://digitalrepository.unm.edu/nma/vol4/iss5/2/.

12 Herbert Bayer, "articulated structures," Herbert Bayer Papers, Denver Public Library.

13 Robert O. Anderson, *Fundamentals of the Petroleum Industry* (University of Oklahoma Press, reprint edition, 1984), 244.

14 Heather Davis, *Plastic Matter* (Duke University Press, 2022), 46.

15 Ibid., 47.

SONJA DÜMPELMANN

SOUNDING EXTREME ENVIRONMENTS

Sonja Dümpelmann is a landscape historian and professor at Ludwig Maximilian University of Munich where she holds the chair of Environmental Humanities. She is the author and editor of several books, most recently of the award-winning *Landscapes for Sport* (Dumbarton Oaks, 2022) and *Seeing Trees: A History of Street Trees in New York City and Berlin* (Yale University Press, 2019). She lectures internationally and has served as senior fellow in Garden and Landscape Studies at the Dumbarton Oaks Research Library and Collection, Washington, DC.

+ LANDSCAPE HISTORY, DESIGN

Seed cones pop, and leaves rustle. Woods muffle sound, as does grassland. Plants make sound. They also re- and deflect, scatter, attenuate, and absorb sound, creating specific acoustic environments. Although our perception and creation of these soundscapes span human history, the physics of sound only began to play an increasing role in landscape and urban design in the post-World War II decades between the 1950s and 1970s.[1] This is also the period when environment became a term defining an era and a movement.[2] Wars were destroying the environment. Industry (including its sounds) was polluting it, and too few governments, it appeared, were protecting it. Much of what we know about the environment and how we know it, is related to crisis events in human history and to colonial projects in extreme environments such as tropical forests, oceans, and outer space. This is also true in the case of "plant acoustics," the study of plants' capacity to effect sound waves and thereby alter the human acoustic environment. Plant acoustics built upon expertise gained for warfare in extreme environments. The civilian relevance of this knowledge became clear once sound in densely populated urban areas and along highways was increasingly perceived as disturbing and unwanted noise. What I call plant acoustics in this article is different from the more recent development of plant bioacoustics, the study of how plants themselves make and respond to sounds.[3] In contrast, plant acoustics was built on an anthropocentric worldview in which plants were seen mostly in terms of their instrumentality for human life and, as in the context presented here, as a "vegetation technology."

After World War II, the growth of private motorization, land and air traffic caused unprecedented sound pollution especially in urban areas in the United States and West Germany, the leading countries in plant acoustics research. The total number of automobiles in the US rose from approximately 40.3 million in 1950 to 59.2 million in 1959.[4] Fueled by postwar reconstruction and the economic miracle years, in West Germany (without West Berlin and Saarland), car ownership grew significantly, from nearly 500,000 in 1950 to 3.5 million in 1959.[5] Increased jet air travel required the enlargement of existing airports and the building of new ones. Many citizens approached their governments with noise grievances. Besides the development of quieter engines, the regulation of traffic, and the establishment of noise ordinances, architectural, landscape, and urban design measures were considered, if not the problem's solution, then a means for its mitigation. Plants appeared to some as a valuable measure in this effort. Given vegetation's other beneficial functions, including its ability to intercept stormwater, provide windbreaks, filter and cool the air, and enhance the urban fabric aesthetically, plants' use to protect against unwanted sound appeared especially fitting and a worthwhile research endeavor.

Setting the Tone

For at least half a century since the early 1900s, cities have developed ways to mitigate the increasing diversity and intensity of sounds caused by industrialization and urbanization. In the early 20th century, they crafted noise ordinances and zoning laws to separate industrial and residential areas and protect inhabitants from sound besides smoke and disturbing smells. Architects and engineers used knowledge in acoustics for the design and construction of concert halls, theaters, and auditoria, and then for the first soundproof buildings; physicians began to study the effects of noise on human health; and citizens rallied in anti-noise campaigns. In New York City–soon infamous as the noisiest city in the United States–philanthropist Julia

Barnett Rice founded the Society for the Suppression of Unnecessary Noise in 1906.[6] In Germany two years later, philosopher Theodor Lessing established a similar, albeit even more elitist and short-lived, Anti-Noise Society (Antilärmverein) to organize against noise.[7]

City building directors and urban designers analyzed the urban environment to propose measures of relief. These mostly regarded the city's inert materials, such as "soundless street paving" (i.e., asphalt and woodblocks) and vehicular components like rubber tires.[8] However, in his influential 1896 handbook on hygiene in urban design, Josef Stübben also recognized plants' sound-muffling effect. According to the German architect, the vegetation of enclosed, gated squares protected against traffic noise, a prerequisite for providing proper "places of recreation" (Erholungsplätze).[9] In keeping with much of the time's anti-noise discourse, vegetation and tranquility were used to create and mark bourgeois space in the city. A step beyond the classist use of vegetation for sound attenuation was the architect Hans Christian Nussbaum's encouragement to plant trees in tenement courtyards so that their foliage would scatter and muffle sound.[10] Martin Wagner, who would serve as Berlin's building director from 1926 until 1933 when the Nazis forced him to retire from office prematurely, mentioned parks' and gardens' ability to attenuate sound in his pathbreaking 1915 dissertation on "The Sanitary Green of Cities," but only in passing.[11] In 1934, the Society of German Engineers' Committee for Noise Reduction (Fachausschuß für Lärmminderung) noted that trees in residential front yards were beneficial due to their sound buffering effect.[12]

In the US, noise measurements undertaken by New York City's Noise Abatement Commission, established in the Department of Health in 1929, sparked interest in plant acoustics. In 1931, urban planner Robert Whitten noted the noise absorption by grasses, shrubs, and tree foliage and encouraged scientific studies to determine how vegetation, in connection with ground modulation, could best be used for sound attenuation in express highway design. After all, the densely growing creeping bentgrass, which began to be widely used on golf courses at the time, could turn out to be a more effective sound absorber than other grass species, Whitten suggested.[13] Ten years later, American landscape architect Ralph Rodney Root offered one of the most extensive discussions of plants' ability to mitigate sound in the first half of the 20th century. His elaborations profited from the knowledge gained in architectural acoustics. Root considered "noise quieting" fundamental in "contourscaping," a neologism he invented to describe landscape design, perhaps under the impression of the Midwestern dust storms and farmers' need to plow along contours. He suggested plantings as noise screens, with evergreen plants being more effective year-round sound mufflers than deciduous species. According to Root, mixed and irregularly planted vegetation could absorb, scatter, and deflect sound waves, for example, toward the ground or away from buildings into the open sky. In contrast, "clipped or sheared plant material and straight even plantings with the same height," as well as street trees with arching branches, produced reverberant conditions, which increased the noise level.[14]

Only studies conducted years later confirmed street trees' tunnel-effect and showed that low frequencies in city streets dissipated into the sky and along the street channels while ivy-covered building walls could muffle sound.[15] Until the mid-20th

1 For the term soundscape and its early conception and use, see R. Murray Schafer, *The Soundscape: Our Sonic Environment and the Tuning of the World* (Destiny Books, 1977).

2 Paul Warde, Libby Robin & Sverker Sörlin, *The Environment: A History of the Idea* (Johns Hopkins Press, 2021).

3 For an engaging overview of the current state of research into phytoacoustics, see Karen Bakker, *The Sounds of Life: How Digital Technology Is Bringing Us Closer to the Worlds of Animals and Plants* (Princeton University Press, 2022), 99–118.

4 See https://www.fhwa.dot.gov/ohim/summary95/section2.html, accessed September 10, 2024.

5 See the statistics for January 1, 1951 and January 1, 1960 in *Tatsachen und Zahlen aus der Kraftverkehrswirtschaft* 1951, no. 15 (1951): 110-111; *Tatsachen und Zahlen aus der Kraftverkehrswirtschaft* 1959/60, no. 24 (1960): 29.

6 Emily Thompson, *The Soundscape of Modernity: Architectural Acoustics and The Culture of Listening in America, 1900–1930* (The MIT Press, 2004), 120–30.

7 Klaus Saul, "Wider die 'Lärmpest'. Lärmkritik und Lärmbekämpfung im Deutschen Kaiserreich," in Dittmar Machule, Olak Mischer & Arnold Sywottek (eds), *Macht Stadt krank? Vom Umgang mit Gesundheit und Krankheit* (Dölling und Galitz, 1996), 154-92.

8 Georg Pinkenburg, *Der Lärm in den Städten und seine Verhinderung* (G. Fischer, 1903).

9 J[osef] Stübben, *Hygiene des Städtebaus* (Gustav Fischer, 1896), 420.

10 H. Chr. Nussbaum, *Das Wohnhaus und seine Hygiene* (Alfred Kröner, 1909), 41.

11 Martin Wagner, "Das sanitäre Grün der Städte," Ph.D. dissertation, Berlin University, 1915, 4.

12 Verein Deutscher Ingenieure, *Das lärmfreie Wohnhaus* (VDI, 1934), 88.

13 Robert Whitten, "The Prevention of Blight in Residential Areas Adjacent to Expressways," *Planning Problems of Town, City and Region, Papers and Discussions at the Twenty-Third National Conference on City Planning held at Rochester, New York, June 22–24, 1931* (Philadelphia, 1931), 164–72.

14 Ralph Rodney Root, *Contourscaping* (Ralph Fletcher Seymore, 1941), 228–35.

15 David F. Van Haverbeke & David I. Cook, "Green Mufflers," *American Forests* 87, no. 11 (1972): 28–31 (30); David I. Cook & David F. Van Haverbeke, "Trees and Shrubs can Curb Noise but with Quite a Few Loud 'Ifs'," in *Yearbook of Agriculture U.S. Department of Agriculture*, ed. by U.S. Department of Agriculture (U.S. Department of Agriculture, 1972), 28–30; David Aylor, Jean-Yves Parlange & Courtland Chapman, "Reverberation in a City Street," *The Journal of the Acoustical Society of America 54*, no. 6 (1973): 1754–56.

century, actual studies on plant acoustics were lacking and comments on plants' sound buffering function remained few and far between. Scientific interest began to build up in the postwar years, sparked by the general increase of noise due to mass motorization, especially diesel engines, and related health threats. Still, in the 1970s, scientists seeking to fill the knowledge gap deplored the general lack of "precise information on [trees' and shrubs'] sound-muffling ability."[16] Plant acoustics as an exact science lagged behind architectural acoustics, which had been informing and forming the construction of concert halls and auditoria since the mid-19th century.[17] Although recognized, plants' sound attenuation appeared too difficult to measure and package into the data of hard science. Whereas the acoustics of interior space had been a subject of interrogation for a while, the sound design of outdoor space had remained a lacuna.[18]

Sounding Extreme Environments

Once scientists began to pay attention to plant acoustics in the postwar years, a series of studies undertaken by American physicist Carl F. Eyring and his team during the Second World War was impossible for them to ignore. The team's investigations, conducted in the tropical forest of the Panama Canal Zone in 1944, were part of war research for the U.S. Department of Defense, and they proved groundbreaking for all plant acoustics research that came after.

Involvement in the war's Pacific theater posed numerous new challenges for the US military. One of them was "jungle warfare." Soldiers were confronted with the thick, tangled vegetation of tropical forests impeding all orientation and complicating enemy detection. The plants not only obstructed sight, but also hindered proper judgment of the location of sound sources. The diverse, layered vegetation reflected, diffracted, and scattered sound waves multiple times. Tasked to shed light on this phenomenon, Eyring and his team set out to make the tropical forest more legible and audible to Western eyes and ears unaccustomed to its extreme characteristics and purportedly bewildering complexity. The exploration of "jungle acoustics" was to bring order into the tangled mass of vegetation, thus providing a means of control over terrain.[19] In the Western imagination, the jungle, adapted from the Hindi and Maratha *jangal* and the Sanscrit term *jangala*, was still a negatively connoted social and cultural construct describing environments considered to be impenetrably chaotic, wild and primitive, uncivilized and uncultivated.

Eyring and his team selected their fieldwork site north of Summit on the Atlantic slope, within fifty miles of Fort Clayton (now closed), to resemble virgin tropical forest vegetation as closely as possible. They built a "listening area" consisting of a slightly elevated square platform mounted with their equipment.[20] To determine the error in judging the sound direction of gunshots, team members positioned themselves on the platform facing outward toward the four cardinal directions. The researchers found that the mistake was smallest when the sound came from a direction near the axis passing through the listeners' two ears. It also decreased with increasing distance from the sound source.[21] In addition, the terrain's vegetation quality had an impact on the loss of sound.[22] For these measurements, the scientists chose five sites with different vegetation. They ultimately associated their auditory measurements with the sites' different visual and physical character traits, which they described in terms of visual and physical penetration characteristics: could a moving white

16 Van Haverbeke & Cook, "Green Mufflers"; David I. Cook & David F. Van Haverbeke, *Trees and Shrubs for Noise Abatement, Research Bulletin 246*, The Forest Service U.S. Department of Agriculture (July 1971), 3.

17 See, e.g., Robert S. Shankland, "Architectural Acoustics," *The Journal of the Acoustical Society of America* 61, no. 2 (1977): 250–54; Emily Thompson, "Dead Rooms and Live Wires: Harvard, Hollywood, and the Deconstruction of Architectural Acoustics, 1900-1930," *Isis* 88, no. 4 (1997): 597–626.

18 Also see the comment by Jerzy Sadowski and Lech Wodzinski, "Lärmbekämpfung in Städten," *Deutsche Architektur* 9, no. 2 (1960): 91–93.

19 Carl F. Eyring, "Jungle Acoustics," *The Journal of the Acoustical Society of America* 18 (1946): 257–70.

20 Ibid.

21 Ibid.

22 Ibid.

23 Ibid.

24 Ibid., 266.

25 "Jungle Acoustics," *The Journal of the Acoustical Society of America* 41, no. 6 (1967): 1556–57.

26 Gunther Lehmann, "Die Einwirkung des Lärms auf den Menschen," *Arbeitsgemeinschaft für Forschung des Landes Nordrhein-Westfalen*, Heft 94 (Köln and Opladen: Westdeutscher Verlag, 1961): 7–39.

27 Hans Werner Bobran, "Städtebauliche Grundlagen des Schallschutzes," *Deutsche Bauzeitung* 101, no. 5 (1967): 375–84 (380); Van Haverbeke & Cook, "Green Mufflers," 29.

28 Rieckenberg, "Zur Frage des Lärmschutzes durch Pflanzungen," *Straße und Autobahn* 9, no. 3 (1958): 92–93.

29 Van Haverbeke & Cook, "Trees and Shrubs can Curb Noise but with quite a few 'Ifs'," 29.

30 Donald Aylor, "Sound Transmission through Vegetation in Relation to Leaf Area Density, Leaf Width, and Breadth of Canopy," *The Journal of the Acoustical Society of America* 51, no. 1 (1972): 412.

31 Donald Aylor, "Noise Reduction by Vegetation and Ground," *The Journal of the Acoustical Society of America* 51, no. 1 (1972): 204–205. Also see D.E. Aylor, "Muffling Noise with Soil and Plants," *Ground Maintenance* (August 1972): 20–22.

32 For the US, see e.g., Cook & Haverbeke, *Trees and Shrubs for Noise Abatement*; David I. Cook, "Discussion: Highway Noise and Acoustical Buffer Zones," *Transportation Engineering Journal of ASCE* 101, no. TE2 (1975): 402–03.

33 Dümpelmann, *Seeing Trees: A History of Street Trees in New York City and Berlin* (Yale University Press, 2019), 181–83; Gerhard Beck, *Pflanzen als Mittel zur Lärmbekämpfung* (Patzer,

object be seen, and at which distance? Was walking in the forest hard or easy, and did it require cutting underbrush? For all studies, a general understanding of sound transmission loss of audible frequencies due to tropical weather phenomena was important. Working during the wet season, they found that while temperature and wind velocity gradients were negligible in the tropical forest, humidity increased the transmission loss of high frequencies.[23] Lastly, the team also identified the sound spectrogram of characteristic ambient "jungle noise," finding that during the day and between animal calls, the ambient sound level was very low, even appearing "deathly silent."[24] At night, however, the volume of low frequencies decreased, whereas that of high frequencies increased due to insects' nocturnal chorus. Eyring's study provided the starting point for a similar investigation some years later at the height of the Vietnam War and for civilian explorations into the use of plants for attenuating sound in urban environments.[25]

Plant Acoustics

In the 1950s, urban areas were quickly becoming ever more extreme environments. Noise pollution triggered an increased sound consciousness among citizens and design professionals. In West Germany, reconstruction activities produced much din, as did the implementation of new urban design paradigms that prioritized automobile transportation, thus increasing traffic noise. In the US, highway construction and suburbanization similarly posed new audible challenges. The rise of sound levels and frequencies during the postwar economic boom was found to cause serious health threats that ranged from psychological reactions to high blood pressure and mechanical injuries in the hearing system.[26] Nevertheless, it was often other environmental or economic grievances that brought home the point of harmful noise: the underground voles that died en masse when a new helicopter type with high sound frequencies was introduced at a West German military airbase and the hens that laid fewer eggs once the tree belt protecting the chicken farm in Kansas from highway noise was felled.[27]

Physicians, urban designers, highway planners, and landscape architects realized the potential of plants in noise abatement, not least given their many other benefits. If vegetation could shield against unwanted sights, why not against sounds? As medical acoustician Franz Joseph Meister posited in 1955, plants were one of the best ways to improve urban space, not only from aesthetic and environmental points of view but in relation to public health. Eyring's "jungle acoustics" quickly became researchers' reference point and remained the foremost reference for decades. Scientists set up their loudspeakers and microphones in various monocultures of evergreen and deciduous forests in North America and Europe, in mixed woods, near plantings along highways, urban streets, and parks. Their sound measurements confirmed the general tendencies of sound mitigation described by Eyring for plants in tropical forests.

One of the challenges regarding plant acoustics research was that it was difficult to undertake in a lab and was dependent upon field studies. These were seldom comparable. Site conditions, technical setup, and study parameters varied, and the site selection relied on existing vegetation that was not planted for the respective study purposes. Inadvertently, large-scale vegetation engineering projects undertaken in North America and on the European continent in the 1930s and

1967); Gerhard Beck, "Untersuchung über Planungsgrundlagen für eine Lärmbekämpfung im Freiraum mit Experimenten zum artspezifischen Lärmminderungsvermögen verschiedener Baum- und Straucharten," Ph.D. dissertation, Institut für Gartenkunst und Landschaftsgestaltung, Technische Universität Berlin, 1965.

34 Ulrich J. Kurze, "Noise Reduction by Barriers," *The Journal of the Acoustical Society of America* 55, no. 3 (1974): 504–18; Lee P. Herrington, "Trees and Acoustics in Urban Areas," *Journal of Forestry* 72, no. 8 (1974): 462–65; L. E. DeFrain, "Highway Noise Abatement by Design?" Proceedings of the Conference on Noise Control, Washington, D.C., October 15–17, 1973, 99–104; Donald E. Aylor & Lawrence E. Marks, "Perception of Noise Transmitted Through Barriers," *Journal of the Acoustical Society of America* 59, no. 2 (1976): 400.

35 Franz Josef Meister & Walter Ruhrberg, "Die Dämmung von Verkehrsgeräuschen durch Grünanlagen," *VDI Zeitschrift* 101, no. 13 (1959): 527–35.

36 White House Conference on Natural Beauty, Washington, D.C., 1965: Beauty for America: proceedings, May 24–25, 1965 (U.S. Government Printing Office, 1965), 85. Sonja Dümpelmann, "Airport, Landscape, Environment," in Sonja Dümpelmann & Charles Waldheim (eds), *Airport Landscape: Urban Ecologies in the Aerial Age* (Harvard Graduate School of Design, 2016), 97–119.

37 Federal Aviation Agency, *Dulles International Airport, Construction Progress Report* (U.S. Government Printing Office, without date), 23; See *Aircraft Noise Problems, Hearings Before Subcommittees of the Committee on Interstate and Foreign Commerce, House of Representatives, Eighty-Sixth and Eighty-Seventh Congress* (U.S. Government Printing Office, 1963), 313, 549, 561.

38 Van Haverbeke & Cook, "Green Mufflers," 31.

39 Hans Werner Bobran, "Städtebauliche Grundlagen des Schallschutzes," *Deutsche Bauzeitung* 101, no. 5 (1967): 380.

40 Franz Josef Meister & Walter Ruhrberg, "Die Dämmung von Verkehrsgeräuschen durch Grünanlagen," *VDI Zeitschrift* 101, no. 13 (1959): 534; F. J. Meister, "Straßenverkehr und Lärm," and "Städtebauliche Abhilfen," in *Medizin und Städtebau: Ein Handbuch für gesundheitlichen Städtebau* vol. 2, ed. by Paul Vogler & Kühn (Urban & Schwarzenberg, 1957), 10–21, 43–47.

41 Hans Bernard Reichow, "Städtebau und Lärmbekämpfung," *Kampf dem Lärm* 9 (1962): 96–104, 130–37.

42 See Floyd Watson, *Acoustics of Buildings* (Wiley, 1923), 25.

43 J. E. Moore, *Design for Noise Reduction* (Architectural Press, 1966), 88, 134; J. E. Moore, *Design for Good Acoustics and Noise Control* (Architectural Press, 1978), 91–92, 116.

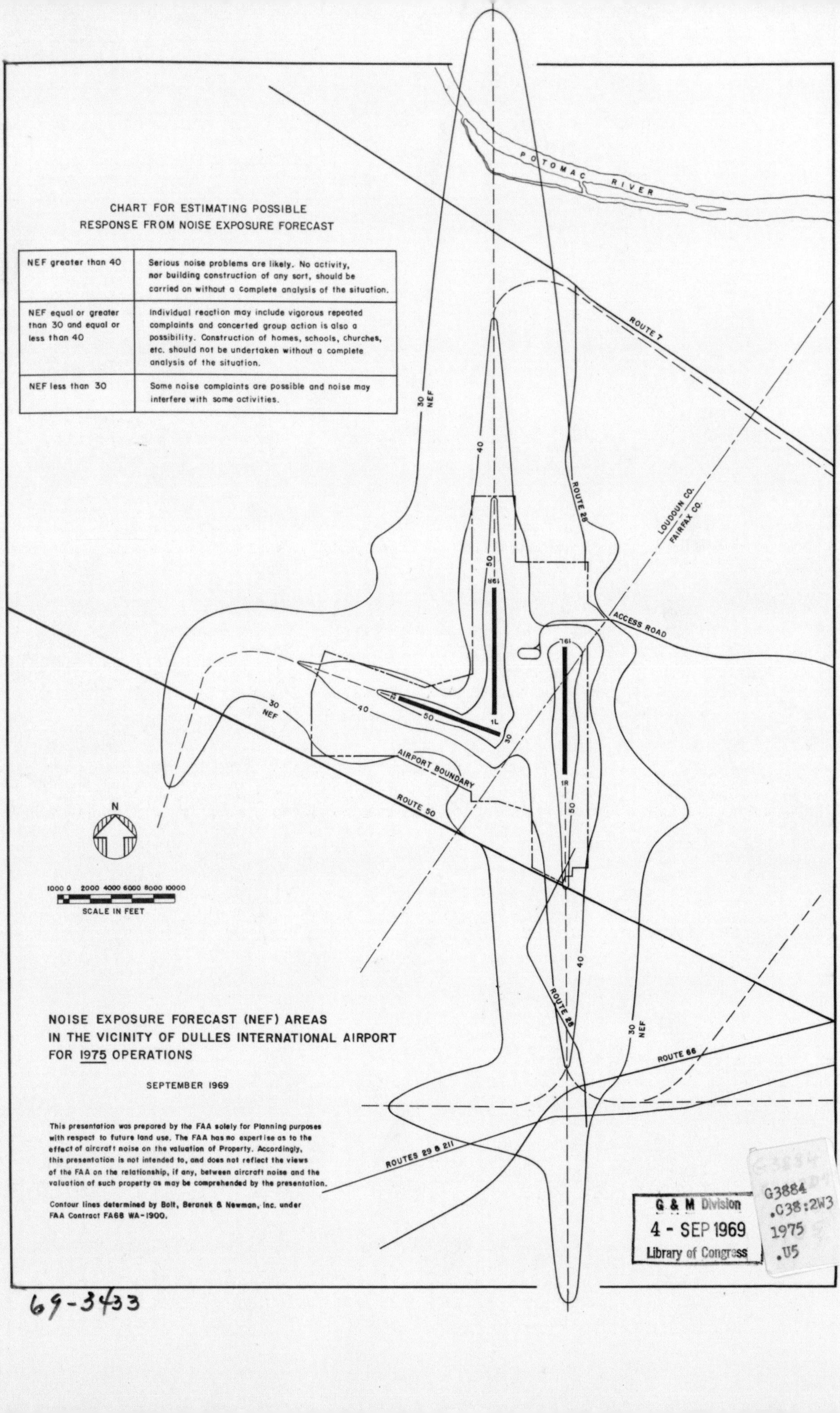

CHART FOR ESTIMATING POSSIBLE
RESPONSE FROM NOISE EXPOSURE FORECAST

NEF greater than 40	Serious noise problems are likely. No activity, nor building construction of any sort, should be carried on without a complete analysis of the situation.
NEF equal or greater than 30 and equal or less than 40	Individual reaction may include vigorous repeated complaints and concerted group action is also a possibility. Construction of homes, schools, churches, etc. should not be undertaken without a complete analysis of the situation.
NEF less than 30	Some noise complaints are possible and noise may interfere with some activities.

NOISE EXPOSURE FORECAST (NEF) AREAS IN THE VICINITY OF DULLES INTERNATIONAL AIRPORT FOR 1975 OPERATIONS

SEPTEMBER 1969

This presentation was prepared by the FAA solely for Planning purposes with respect to future land use. The FAA has no expertise as to the effect of aircraft noise on the valuation of Property. Accordingly, this presentation is not intended to, and does not reflect the views of the FAA on the relationship, if any, between aircraft noise and the valuation of such property as may be comprehended by the presentation.

Contour lines determined by Bolt, Beranek & Newman, Inc. under FAA Contract FA68 WA-1900.

1940s turned out to be practical for the purpose. In 1955, German scientists used an autobahn buffer planting from 1938 to measure the sound muffling effect, which, however, appeared negligible in the 1955 study, thus sparking further research elsewhere.[28] In the Midwestern US, the shelterbelts planted against wind and topsoil erosion for the Prairie States Forestry Project in the 1930s and 1940s provided welcome test sites.[29] In one of the two most extensive studies carried out in the US in the late 1960s and early 1970s, several shelterbelts planted during the "Dust Bowl" days in relatively quiet locations northwest of Lincoln, Nebraska, were used. Prerecorded sounds of highway trucks, arterial passenger cars, and bus stops were blasted through the vegetation, and, at various distances behind the shelterbelts, the reduced sound was measured. Measurements were compared to those obtained at the same distances on similar terrain that lacked shelterbelts.

Based on such studies, researchers determined that densely planted strips of trees and shrubs, 20 to 50 feet wide, appeared to be the simple gold standard for muffling noise along urban roads. The wider, thicker, and more numerous the leaves, and the denser the foliage, the more effective the plantings were in sound attenuation. Several tests also showed that berms, embankments, or sinking roadways could further enhance the plants' muffling effects. Most studies conducted in the US and the two German states between the 1950s and 1970s found that trees and shrubs were most effective at attenuating high frequencies. While an American research team concluded that foliage, stems, and ground conditions worked in concert when attenuating sound, they also proposed that the design of "vegetative mufflers" should respond to the sound frequency to be mitigated.[30] Their studies had revealed that a strip of soft ground was most effective in combatting the low-frequency noise of trucks starting up at toll booths, but a band of dense foliage was best to reduce the noise from high-speed traffic.[31]

Whereas field studies in the expansive US were primarily undertaken in rural and suburban areas, research in densely settled Europe concentrated on urban areas. On the North American continent, large-scale plantings dominated, and scientists identified and noted the role of their more general characteristics in sound attenuation, such as deciduous versus evergreen foliage, tree height, tree belt depth, and density.[32] Researchers in Europe, on the other hand, homed in on the details. In the early 1960s, landscape researcher Gerhard Beck identified a lack of knowledge concerning sound attenuation vis-à-vis different species characteristics. These included leaf size, form, position, and texture; foliage density and distribution; as well as the character of trunks, branches, and twigs. In his dissertation and postdoctoral work at the Technical University Berlin, one of the two centers of plant acoustics research in West Germany, Beck demonstrated that trees and shrubs' sound attenuation was species-specific. Particularly effective were the large and hard-leafed species, species with leaves oriented perpendicular to the sound waves, and those with many dry interior leaves and branches. Sycamores and little-leaf linden, for example, turned out to have an especially big buffer effect during the summer months.[33]

Overall, research found plants to be influential sound mufflers, depending on their properties and especially in conjunction with their psychological effects. Studies showed, for example, that regardless of their actual sound muffling effect, plants

Opposite: A 1969 map of "Noise Exposure Forecast Areas in the vicinity of Dulles International Airport for 1975 operations."

could have an impact on perceived loudness. Human subjects perceived of plants that did not hide the sound source entirely as more successful sound mufflers than vegetation that fully obscured the sound source.[34]

Planting Acoustic Environments

To facilitate the application of sound research in landscape design, early acoustics researchers Franz Josef Meister and Walter Ruhrberg made recommendations using references that many landscape architects would have been familiar with. The researchers described the more effective buffer plantings as resembling bosquets in French gardens during the period of Louis XIV. They used a regional reference to 19th-century landscape gardens designed by court gardener Maximilian Weyhe to explain that in loose naturalistic plantings, it was as much the spaciousness as the materiality that created sound guards, albeit less effective ones.[35]

Deliberate efforts to mitigate noise through planting design and ground modulation were undertaken in planning new residential neighborhoods, highways, roadways, and airports. Often, designs consciously or unconsciously conflated both plants' visual and sound shielding functions. Among the earliest attempts to use plants for large-scale noise protection was the planting of "hush parks" near jet airports in the US.[36] Dulles and Minneapolis–St. Paul International Airports are two examples where sufficient land for ground modulation and expansive plantings to alleviate noise from landing planes and engine runups before take-off was available. At Dulles, the Federal Aviation Agency initiated a reforestation program to ensure a solid 300-meter-wide tree belt along the airport perimeter that would be widened even further to 760 meters at the ends of the runways.[37] At the Twin Cities' Wold-Chamberlain Airport, a park-like green belt was laid out so that its plantings and topography would protect adjacent residential subdivisions against aircraft noise.[38]

In Germany, planners suggested the establishment of sound protection zones on the scale of the Federal Republic.[39] However, noise protection was also considered on the scale of individual neighborhoods and new towns. The number of street crossings was to be reduced to enhance traffic flow, and residential buildings were to be positioned so that they both shielded and were shielded from noise. Plants and planting designs played an important role in these concepts, although making specific recommendations for landscape design was more difficult than for traffic planning and overall site design. Based on their research Meister and Ruhrberg offered two prototypical designs for "sound-proof" neighborhoods "organically embedded into the landscape."[40] Among those lobbying for sound-conscious site design was architect Hans Bernhard Reichow, who during the Second World War had supported the Nazi's expansionist planning and reconstruction projects, aspects of which after the war continued to inform his "organic city building" and car-friendly city paradigm. Reichow sought to practice what he preached in his plans for a small residential town near Schwalbach, north of Frankfurt, and for the model new town Sennestadt near Bielefeld.[41]

Despite the evidence, some scientists' credence and confidence in using plants to muffle noise was low. In their eyes, plants' individuality, relative unpredictability, and seasonal and life cycles disqualified them for sound attenuation. When, in the early 20th century, architectural acoustics pioneer Wallace C. Sabine and his colleagues measured the sound absorption of various materials and objects in interior spaces, house plants scored lowest.[42] In keeping with this, about half a century later, a widely used handbook on architectural acoustics described the outdoor plant remedy against noise as over-optimistic. Author John Edwin Moore conceded that wide vegetation screens could have a notable effect, especially at high sound frequencies. In general, however, he warned that plants' sound absorption was relatively small, primarily due to their foliage, making it unpredictable.[43] Theodore John Schultz, the technical director of architectural acoustics and noise control at the American research and development company Bolt, Beranek, and Newman (BBN), even ridiculed the efforts in plant acoustics research as perpetuating "an unfortunate bit of folklore"; and this although his firm had been among the first to test (and confirm) Eyring's results in the context of the northeastern forests in the US. According to Schultz, plants' crucial fault was that they could not be installed with immediate effect and would not provide the aspired density and extent necessary for effective sound attenuation. Much more successful and economical were fence-type barriers, Schultz argued.[44] He was not writing without self-interest. By the 1970s, BBN was a pioneer in designing highway noise barriers besides the country's prime computer modeler of roadway and aircraft noise and a consultant on all things acoustic.[45]

Although vegetation is a significant characteristic of many environments on this planet, plant acoustics has only occupied a niche in the scientific study of sound. One of the reasons why less attention has been paid to plant acoustics than to architectural acoustics is because it is complicated by the fact that plants are alive. Their sound attenuation effects change with their seasonal and life cycles. They grow in myriad forms and shapes, even within the same species or cultivar, and their forms change with time. Plants cannot be specified and reproduced the same way inert materials can. Their inherent life force may also pose ethical questions, especially since plant bioacoustics has shown more clearly that plants sense and emit sound waves.[46] As for humans and other animals, we may also need to ask, how much noise is too much for plants to live healthy and productive lives, or which sounds enhance their growth? At the bottom of this lies the question of how landscapes can be designed to serve more than human needs. In the 1960s, landscape researcher Beck saw in the study of plants' sound mitigation a means to "humanize" the environment, making it more livable for humans.[47] Perhaps knowledge in both plant acoustics and plant bioacoustics could lead us to design more livable environments for life on earth in general.

44 Theodore John Schultz, "Discussion: Highway Noise and Acoustical Buffer Zones," *Transportation Engineering Journal of ASCE* 101, no. TE2 (1975): 403–4.

45 The firm had served as acoustical consultant to communities, airport authorities, and the Federal Aviation Administration from the moment jet aircraft noise was recognized as a nuisance in the early 1950s. See *Aircraft Noise Problems, Hearings Before Subcommittees of the Committee on Interstate and Foreign Commerce, House of Representatives, Eighty-Sixth and Eighty-Seventh Congresses* (U.S. Government Printing Office, 1963) and Leo Beranek, *Riding the Waves: A Life in Sound, Science, and Industry* (The MIT Press, 2008).

46 Monica Gagliano, Stefano Mancuso & Daniel Robert, "Towards Understanding Plant Bioacoustics," *Trends in Plant Science* 17, no. 6 (2012): 323–25; Monica Gagliano, et al., "Acoustic and Magnetic Communication in Plants: Is it Possible?," *Plant Signal Behavior* 7, no. 10 (2012): 1346–8.

47 Gerhard Beck, "Untersuchung über Planungsgrundlagen für eine Lärmbekämpfung im Freiraum mit Experimenten zum artspezifischen Lärmminderungsvermögen verschiedener Baum- und Straucharten," Ph.D. dissertation, Institut für Gartenkunst und Landschaftsgestaltung, Technische Universität Berlin, 1965, 248.

NUCLEAR WEAPONS
AND THE UNITY OF THE ENVIRONMENT

MARCUS OWENS + CHRISTINA ANTIPORDA

Marcus Owens and Christina Antiporda co-founded CAMO Studio in the San Francisco Bay Area as an outlet for research and design-build work in their communities, particularly with Kultivate Labs in SOMA. CAMO interests include the transformation of military lands into recreation areas, cross-border issues, religious landscapes, and drone deployment. Most recently, they taught at Washington State University in Pullman and investigated the management of hot springs on federal lands.

+ DESIGN, TECHNOLOGY

When it comes to the evolution of the word *environment* over the course of the 20th century, its varied, and at times paradoxical, connotations eventually became subsumed into an all-encompassing notion when the invention of nuclear weapons circumscribed the entire planet within a singular fate.[1] The first view of an image of our planet from outer space had a powerful effect on society, and the environmental imagination was similarly shaped by the aesthetic qualities of nuclear weapons, specifically the ways in which their detonation made radioactivity explicit. This collective notion, which registered as a reaction of shock and awe, overpowered previous localized relationships and ideas of the environment. This essay examines the development and deployment of nuclear weapons and their profound impact on both physical landscapes and the broader conceptualization of the environment from the perspective of landscape architecture.

Historian David Nye articulates the aesthetic response to images of nuclear detonations, encapsulating a transition from experiencing natural phenomena as sublime to witnessing human-made phenomena that are equally, if not more, overwhelming. Images of nuclear explosions turn abstract scientific concepts (like radioactivity and nuclear fission) into a tangible, devastating reality, highlighting the catastrophic potential of human ingenuity.[2] The technological sublime associated with nuclear explosions embodies a dual nature – awe at the sheer scale and power of the technology, paired with terror over its destructive capabilities. This dual response is a hallmark of the technological sublime and is central to a concept of environment that unfolded in the nuclear age.

Alongside the phenomenological approach to nuclear detonations, the idea of a "nuclear landscape" also entailed a new way of seeing the Earth that emerged in the post-war techno-scientific planetary view and was highly influential for the practice of landscape architecture. The terror and awe response to satellite and whole Earth imagery likely fueled landscape architect Ian McHarg's use of narratives of space travel and vision in *Design with Nature*, which has led generations of landscape architects and students in both research-driven design and the turn toward social good.[3] The physical landscape is paired with satellite use, and its innovations in communicating, monitoring, and measuring, enable an intimacy with place from remote locations. In this respect, we can identify a pattern in which attitudes toward nuclear weaponry and its global reach pave the way for broader, planetary atmospheric monitoring and actions and debates, such as addressing contemporary climate challenges through climate engineering. This pattern–terror or shock, awe, recovery, and revitalization–reveals itself on multiple scales by the agencies and practitioners who are key to the nuclear era.

Zooming in, narratives of nuclear weapons development suggest a view toward altering environmental conditions related closely to the body. For philosopher Peter Sloterdijk, the 20th century began on April 22, 1915, when terrorism, design, and the environment were linked together through the release of chlorine gas by the German military on French-Canadian troops at Ypres Salient.[4] With the introduction of the environment into the battle between adversaries, the military target became less the body of the soldier than the basic prerequisites for life. Terror, in this sense, consisted of an operational knowledge of elimination that assaulted the environmental conditions required by human life. Design is implicated through the operational manipulation of gas clouds, necessitating certain "non-objective" criteria, such as air quality, light, and temperature. The components and quality of the landscape, such as water, pollution, and the presence of insects or bacteria, become material with which to harm. Design of the environment or environmental technologies poses opportunities for active warfare and anticipates persistent harm to civilian populations. To this extent, we can identify the formation of a "risk society," as described by sociologist Ulrich Beck, in which modern advancements lead to new categories of systemic risks and uncertainties.[5]

Zooming out to a territorial landscape scale, nuclear weapons development required a great deal of physical space, with each site carefully chosen for a unique combination of natural features and particular proximity to, or distance from, urban areas. We also know that the harvesting of materials for chemical warfare leads to long-term effects on the mining sites and the human and nonhuman settlements and activities involved with and around the sites. The afterlives of the lands used for mining, detonations, and other activities are now held by the Department of Energy Office of Legacy Management. This office is focused on tasks of recovery and aspirations toward revitalization.

For Sloterdijk, each step from chemical warfare to nuclear brinkmanship–and into more recent debates about "atmoterrorism," from climate politics to COVID–illustrates a progressive escalation in the scope and depth of environmental manipulation for political and military purposes. Thus, to speak of the environment is not just about conservation or stewardship; it is also profoundly entangled with 20th-century historical practices of environmental control and modification. Design critic and landscape architect Elizabeth Meyer discusses these practices in relation to Beck's risk society as well as his writings "illuminating the degree to which environmental policy is bound to the hazards acceptable within industrial processes."[6] These environmental practices then enmesh with economics, in this case, nuclear energy-related research and development. As Meyer describes:

> Disturbed sites are the byproducts of economic policies that viewed nature as a resource and that accepted environmental degradation as the inevitable consequence of technological progress. The experience of designed landscapes on and in disturbed sites can render visible the consequences of the economic, political, and social decisions that led to those risks.[7]

Meyer summarizes her discussion of designing in this context as an opportunity to express beauty within risk.[8]

Toward this visibility, at the intersection of ecological science and aesthetic representation, landscape architecture confronted the atomic dilemma, grappling with the existential threat of nuclear annihilation while seeking harmonious environmental integration. The 1986 ASLA meeting in San Francisco crystallized this juncture, convening luminaries like Ian McHarg, Karl Linn, and Lawrence Halprin. Their discourse, situated at the twilight of the Cold War, navigated the complex terrains of fear and hope, community and isolation, reflecting on the role of landscape architecture in a world overshadowed by the specter of nuclear conflict. Much emphasis was placed on the spiritual and healing qualities of landscape that could be harnessed and directed in the aftermath of the arms race. The conclusions of the 1986 conference, also evidenced in the trajectory of Linn's career, emphasize the

1 Paul Warde, Sverker Sörlin & Libby Robin, *The Environment: A History of the Idea* (Johns Hopkins University Press, 2021).

2 David E. Nye, *American Technological Sublime* (MIT Press, 1994), 227–28.

3 Ian McHarg, *Design Within Nature* (Natural History Press, 1969), 98–101.

4 Peter Sloterdijk, *Terror from the Air*, trans. Amy Patton & Steve Corcoran (Semiotext, 2009), 18–19.

5 Ulrich Beck, *Risk Society: Towards a New Modernity* (Sage, 1992), 19–30.

6 Elizabeth K. Meyer, "Uncertain Parks: Disturbed Sites, Citizens, and Risk Society,"in J. Czerniak & G. Hargreaves (eds), *Large Parks* (Princeton Architectural Press, 2007), 66.

7 Ibid., 59–86.

8 Ibid.

9 The role of the German nuclear program has been the subject of continued debate and is beyond the scope of this paper, but a few points are worth noting. When the Third Reich invaded Poland on September 1, 1939, the German military focused on the development of reactors and fissile material. Werner Heisenberg reported to Speer and Hitler in 1939 that a bomb was still far off.

10 Richard G. Hewlett & Oscar E. Anderson, Jr, *The New World, 1939/1946: Volume 1 of a History of the United States Atomic Energy Commission* (Pennsylvania State University Press, 1962), 15–17.

11 Robert Alvarez, "Uranium Mining and the U.S. Nuclear Weapons Program," *Federation of American Scientists* (November 14, 2013), https://fas.org/publication/uranium-mining-u-s-nuclear-weapons-program-3/.

12 In fact, it was primarily emigree scientists in the United States, many with communist leanings, who urged Roosevelt to pursue the development of a bomb.

13 General Leslie M. Groves, *Now It Can Be Told: The Story of the Manhattan Project* (Da Capo Press, 2009).

14 Neil Shafer Oatsvall, *Atomic Environments: Nuclear Technologies, the Natural World, and Policymaking, 1945–1960* (University of Alabama Press, 2023).

15 Vannevar Bush, *Science, the Endless Frontier: A Report to the President* (Washington, DC, National Science Foundation, 1945; repr. 1960), https://ia801304.us.archive.org/29/items/scienceendlessfr00unit/scienceendlessfr00unit.pdf.

16 Chunglin Kwa, "Radiation Ecology, Systems Ecology and the Management of the Environment," in Michael Shortland (ed.), *Science and Nature, Stanford in the Vale: British Society for the History of Science* Monograph 8 (BSHS, 1993), 213–49.

(potential) global reach of landscape practice under a unified environmental attitude that encouraged a fusion of landscape practice with social concerns and operation under a pattern of shock, awe, recovery, and revitalization.

Grounding the Nuclear Landscape

A brief timeline of nuclear weapons development in the United States and Germany and the modern techno-scientific apparatus that was used in attempts to usher in this unity of the environment may be useful. The origins of the nuclear bomb can be traced to evidence of nuclear fission published by Otto Hahn and Fritz Strassman in the January 1939 issue of the German nature and science journal *Die Naturwissenschaften*. In the spring of that year, a Uranium Club was formed under the German Army Ordnance Office, primarily focused on producing a nuclear reactor.[9] In August 1939, Albert Einstein and Leo Szilard sent a letter to US President Franklin D. Roosevelt concerning the development of nuclear weapons, prompting the formation of the Advisory Committee on Uranium.[10] By 1940, this committee had been reorganized and expanded into the Uranium Committee, part of the US National Defense Research Committee. It explored the feasibility of nuclear weaponry and began to fund larger-scale uranium enrichment and plutonium production experiments at key institutions like Columbia University and the University of Chicago. Uranium mining occurred domestically to supplement material from the Belgian Congo and Canada.[11] This is the initial grounding of the nuclear landscape.

The following year, the Uranium Committee was folded into the newly created Office of Scientific Research and Development (OSRD), established in June 1941 by Roosevelt and directed by Vannevar Bush. Under the OSRD, the scope and resources of the project significantly increased, which consolidated and centralized the US government's nuclear research efforts into a formalized, large-scale operation dedicated to the rapid development of atomic weapons.[12]

Despite the ongoing public debate over isolationism, the OSRD had already begun concentrated efforts to develop atomic weapons before the United States entered the war against Germany on December 11, 1941. In August 1942, the US Army Corps of Engineers (USACE) formally established the Manhattan Engineer District. This new

organization was specifically tasked with the development and production of atomic weapons, and it effectively took over all the prior research and development efforts that had been managed by the Uranium Committee under the OSRD. The district's primary research and production facilities to produce nuclear weapons were established in Oak Ridge, TN, Los Alamos, NM, and Hanford, WA, by January 1943.[13] Today, each of these sites forms a unit of the Manhattan Project National Historic Park.

Weapons, Energy, Environment

After World War II, in 1946, the US government transitioned the administrative responsibilities of the wartime Manhattan Project into a civilian agency known as the Atomic Energy Commission (AEC). By 1947, all other OSRD contracts and responsibilities had also been terminated or transitioned. The AEC was tasked with regulating nuclear energy and weapons materials, conducting nuclear research, and overseeing the peaceful and military applications of atomic energy.[14] In 1974, responding to the growing importance of energy beyond nuclear power and the need for a more comprehensive energy policy following the 1973 oil crisis, the government restructured the AEC into the Energy Research and Development Administration (ERDA). Today's Department of Energy (DOE) was established in 1977 by consolidating ERDA and various other energy-related agencies to form a single cabinet-level department dedicated to managing national energy policies and programs, including energy conservation, energy research, and the nuclear weapons program. Accordingly, the department became administrator of vast land holdings for these purposes.

In addition to the geopolitical context and diverse spatial contexts in which the Manhattan Project operated, the conceptual framework that grew out of this program continued to shape 20th-century approaches toward the environment. Alumni from the Manhattan Project sought a model for structured research, rigorous funding mechanisms, and centralized coordination to achieve substantial technological advancements.[15] In the aftermath of World War II, these principles were institutionalized through the creation of the National Science Foundation (NSF) in 1950. As laid out by Bush, the NSF was designed to support fundamental research and education in all the non-medical fields of science and engineering. It adopted a peer-review process as a cornerstone for funding decisions, ensuring that investments in scientific

Above and Opposite: Weldon Spring Site radioactive waste disposal cell, St Louis, Missouri.

research were based on potential for impact. The legacy of the Manhattan Project, thus, extends beyond its immediate military outcomes, influencing the structure and methodology of scientific inquiry that defines modern technoscientific expertise. With the NSF, we can begin to see the clarification of a nuclear-bounded concept of the environment.

Perhaps the clearest link between the Manhattan Project and modern environmental research is in Howard T. Odum's work.[16] His involvement with the Manhattan Project significantly influenced his pioneering contributions to ecology, particularly in the areas of systems ecology and ecological modeling. His work during this period, which involved assessing the environmental impacts of uranium mining and nuclear weapon use, informed his later theories about energy flows and systems dynamics within ecosystems. Odum's concepts of energy flow through ecosystems and ecological efficiency became cornerstones of ecological science, emphasizing the importance of viewing ecosystems as complex, interconnected systems. This systems-oriented approach led to advances in ecological modeling, which have been crucial for environmental management and sustainability practices.

Following new environmental research by Odum and others, new legislation began to regulate the spatial production of nuclear weapons. The Manhattan Project utilized an array of facilities across the United States, from the primary facilities that now make up the Manhattan Project National Historical Park to more mundane storage warehouses in New York City and processing facilities in Buffalo and St. Louis, demonstrating a broad geographical distribution that was essential for its secretive and expansive operations.[17] In addition to these facilities, the raw materials for the scientific work included uranium ore mined on the Colorado Plateau, with contracts requiring the inclusion of the Diné (Navajo) on whose reservations the mines were located.[18] In the wake of the National Environmental Policy Act of 1969 and the formation of the Environmental Protection Agency in 1970, programs such as the National Environmental Research Park (1972)—carved out of federally protected lands and utilized as outdoor laboratories—as well as the Formerly Utilized Sites Remedial Action Program (FUSRAP) of 1974 began to address environmental contamination from America's early atomic energy programs.[19] Further legislation, such as the Comprehensive Environmental Response, Compensation, and Liability Act (CERCLA, 1980), more commonly known as Superfund, enabled the significant rehabilitation of landscape holdings. Program sites—the DOE calls them ecosystem sanctuaries—include the three units of the Manhattan Project National Historic Park, Weldon Spring, and Fernald Preserve.

Along with classifying and cleaning up Manhattan Project sites, over the course of the 1970s, the AEC (and then the DOE) began to identify locations for long-term nuclear waste disposal. Carlsbad, NM, was identified as a site for a Waste Isolation Pilot Plant. America's only deep geological repository currently in operation for the disposal of radioactive waste was chosen for its thick salt formations, which naturally self-seal and effectively isolate the waste by slowly encasing it in salt. Between the plans for the National Environmental Research Parks, FUSRAP, and the new Waste Isolation Pilot Plant, the DOE stepped more fully into its role as a land management agency, distancing itself from the spectacle of detonations tourism and toward the recovery and revitalization of land.

Futures Communication

Soon after the Carlsbad site was chosen for long-term waste storage, the DOE contracted a multidisciplinary Human Interference Task Force to produce a series of reports over the 1980s and into the early 1990s. The task force was established with a unique and forward-thinking mission: to design communication and marking systems to deter human interference with nuclear waste repositories for up to 10,000 years. This multidisciplinary group collectively grappled with the profound challenge of conceiving messages that could transcend eons – effectively communicating with future civilizations.[20] This timespan and the assumed rate of decay of the radioactive materials give the nuclear landscape a temporal scale. The task force had to consider how future humans might perceive and interact with these markers without any guarantee that contemporary languages, symbols, or cultural contexts would still be understood.[21] Thus, they explored how to encode messages about the dangers of buried waste in ways that might be universally surmised. This included the use of threatening earthworks, forbidding landscapes, and other elements that instinctively signal danger; in this, we recognize a striving for the technological sublime.

It is a challenge that continues to intrigue designers across the globe today, as evidenced by the 2017 Arch Out Loud competition titled "Nuclear: Landmarker for a Waste Isolation Site," the 2015 competition organized by the French nuclear waste agency, Andra, called "Conserver et Transmettre la mémoire," and the Spring 2024 landscape architecture studio at Washington State University, "Nuclear Futures." The 2017 winning proposal, "Testbed," by New York-based Agency-Agency, eschewed ancient or cartoonish references and proffered chemical reactions orchestrated for landscape spectacle and recognizable as unnatural and intentionally repellent.[22] For the Task Force and competition entrants, designs for these sites expanded the horizon of the "long-now" into "futures" or "artificial" archeological sites.[23] This is to say, it is not an analog of a functional architectural assemblage but a reproduction of monumental ruins that have endured and been discovered by future humans after a process of decay. In this sense, the notion of designing for "deep time" is noble but ultimately Sisyphean. Perhaps this is why most nuclear waste disposal projects initiated in the 1990s utilized the task force's more moderate and passive communications approaches: legal documentation, mapping, and even oral histories.[24]

Meanwhile, more standard venues of communication continued to host landscape discourse, such as the ASLA Conference and publications. However, reflecting on the 1986 ASLA conference in San Francisco, we feel that the moderation and passivity in design outcomes at FUSRAP and other nuclear sites miss the urgency that Karl Linn and a group of landscape architects then identified (though most are touted by the DOE as success stories). At the time, the nuclear arms race was of primary and urgent concern. Cheryl Barton, the ASLA president-elect at the time, stated, "It is a newly established goal of [ASLA] to take a public stand on issues of major national and global significance."[25] Linn, per his introduction at the conference, had studied landscape architecture and psychology, practiced and professed landscape architecture, and had by 1986 retired from teaching but redirected his energies to "peace and social responsibility issues."[26] This included participation in the collective Architects/Designers/Planners for Social Responsibility, which has a contemporary life as a nonprofit policy change advocacy group.[27] For the conference, he called on voices with experience as varied as that of Apollo XIV astronaut Edgar D. Mitchell and Assistant Secretary-General of the UN Robert Muller to address his peers. Though these personages did not appear at the conference, it is clear that Linn felt that landscape architects needed to expand their circles of conversation to address the issues that Barton implies.

This conference took place between two other summits directed at shaping this evolution, each of the three occurring in a crisis context: unrestrained and exploitative development, the Cold War arms race, and the climate crisis.[28] In 1966, McHarg was one of a group of authors to pen the treatise "The Declaration of Concern" naming the "[maintenance of] the vital connection between man and nature" as of the utmost importance to the practice of landscape architecture. The 2016 Landscape Architecture Foundation's *New Landscape Declaration*–itself a reimagining of the 1966 Declaration of Concern–highlighted the development of form and function "to the ideals of equity,

17 US Department of Energy, *Linking Legacies: Connecting the Cold War Nuclear Weapons Production Processes to Their Environmental Consequences* (Office of Environmental Management, 1997); Marco Kaltofen, Robert Alvarez & Lucas W. Hixson, "Forensic Microanalysis of Manhattan Project Legacy Radioactive Wastes in St. Louis, MO," *Applied Radiation and Isotopes* 136 (2018): 143–49.

18 Bindu Pannikkar, Esther Yazzie-Lewis & Doug Brugge, "Ethics of Uranium Mining Research and the Navajo People," in Dianne Quigley, Amy Lowman & Steve Wing (eds), *Tortured Science: Health Studies, Ethics and Nuclear Weapons in the United States* (Routledge, 2017), 143–67.

19 Osti.gov, "National Environmental Research Parks," (July 1, 1994), https://doi.org/10.2172/10161197; C. R. H. Shearer & Nat B. Frazer, "The National Environmental Research Park: A New Model for Federal Land Use," *Natural Resources & Environment* 12, no. 1 (1997): 46.

20 Stephen C. Hora, Detlof von Winterfeldt & Kathleen M. Trauth, "Expert Judgement on Inadvertent Human Intrusion Into the Waste Isolation Pilot Plant," No. SAND90–3063 (1991), https://www.osti.gov/servlets/purl/5789311/.

21 Peter Galison, "Underground Future," in M. Mostafavi & G. Doherty (eds), *Ecological Urbanism* (Lars Muller Publishers, 2010): 304–5.

22 Ella Thorns, "Winners of Landmark for Nuclear Waste Isolation Announced," *ArchDaily*, https://www.archidaily.com/884756/winners-of-landmark-for-nuclear-waste-isolation-announced.

23 Rosemary A. Joyce, "Failure? An Archaeology of the Architecture of Nuclear Waste Containment," in M. Bille & T. F. Sorensen (eds), *Elements of Architecture Assembling Archaeology, Atmosphere and the Performance of Building Spaces* (Routledge, 2016), 424–38.

24 Battelle Memorial Inst., Human Interference Task Force, "Reducing the Likelihood of Future Human Activities That Could Affect Geologic High-Level Waste Repositories" (May 1984), https://www.osti.gov/servlets/purl/6799619.

25 Karl Linn, "The Emerging Landscape of Peace: Proceedings from Three Educational Sessions at the 1986 Annual Meeting of the American Society of Landscape Architects San Francisco," American Society of Landscape Architects and Architects/Designers/Planners for Social Responsibility (1986).

26 Ibid.

27 Raphael Sperry, "Architects/Designers/Planners for Social Responsibility," *Now What?! Advocacy, Activism and Alliances in American Architecture since 1968* (November 27, 2018) https://nowwhat-architexx.org/articles/2018/11/27/architectsdesignersplanners-for-social-responsibility.

28 Elizabeth K. Meyer, "Text and Context: The 1966 Declaration of Concern," in *The New Landscape Declaration: A Call to Action for the Twenty-First Century* (Vireo Books, 2018), Chapter 2.

sustainability, resiliency, and democracy." Though the 1986 conference transcript circulates less than these celebrated statements, its risk society context stands out in visceral association with the technological sublime. Rather than repeat the ideals of 1966, religion, science, and spirituality are nominated as components of an evolving profession. Accordingly, in the existential moment, Linn concludes with the idea of life, calling for a "lifelong commitment" to become "global gardeners." The 1966 document and 1986 talks stress the agency available to landscape architects were they to collaborate and connect more outside of the discipline. The 2016 pledge approaches the attitude Barton, Linn, and ADPSR vociferated but lacks the multidisciplinary bent that the nuclear moment stimulated in landscape architecture conversations.[29]

These theoretical discussions are paralleled by practical achievements in landscape architecture, ranging from small-scale artworks and installations to large parks on formerly industrial sites; between these emerged novel typologies championed by Linn, such as peace parks and community gardens formed in reaction to social conditions. Peace parks have evolved to encompass territorial and international border-spanning wildlife preserves, but in the 1980s, they served as tributes to peace-facilitating efforts. Linn led a workshop on one planned for Crissy Field during the San Francisco conference. He was also involved in an installation called Peace Garden, shown at Fort Mason, with Jeffrey Miller and John Northmore Roberts.[30] At the large park scale, several sites managed under FUSRAP have been transformed. These more recent interventions into legacy management of nuclear sites tend to emphasize the importance of these projects in fostering a deeper public understanding of both the environmental impacts of industrial activities and the potential for landscape architecture to mediate between past harms and future possibilities. However, very little shock and awe remain in these examples. There is a veneer of amnesia, as described by Meyer in her discussion of large parks, in that much of the awe-inspiring history, though perhaps presented on exhibition boards within the new buildings, has been edited out, deemphasized or obscured in the landscape.

As a land manager, the DOE lacked the spectacular vision first outlined by the Human Interference Task Force. The edited design outcomes, with their emphasis on restored native plant communities and little else, present Beck's everyday nature but may lack the dramatic imagery of the technological sublime.[31] For instance, Fernald Preserve in Ohio, a former site for uranium processing, has been redeveloped into a large nature preserve. Today, it offers more than 140 acres of wetlands, 400 acres of forests, and seven miles of hiking trails, along with an on-site visitor center that educates the public about the site's historical significance and extensive cleanup efforts. Similarly, the Weldon Spring Site Interpretive Center in Missouri, which processed uranium and thorium, now features a detailed interpretive center that explains the site's role in the Manhattan Project and the Cold War. The landscape includes a 75-acre disposal cell capped with rock and native grasses, with an accessible viewing platform.[32] Landscape architects are absent from the primary efforts toward restoration and establishing prairie systems in these cases.[33] The design strategy at both restoration sites emphasized ease of maintenance, erosion control, and drought tolerance. At Weldon Spring, site project manager Marj Oaks anticipated that the newly established prairie would soon draw visitors for the "spectacular site of a prairie fire" should a controlled burn be induced in the maintenance regimen for the reintroduced ecosystem.[34] This may be the closest future event to rival the shock and awe of the site's connection with nuclear development.

In conclusion, we can see how the planetary scale terror of nuclear weapons engendered a unifying concept of the environment. For landscape architects and others, this inspired an effort toward outreach and output that could transcend time. However, further interrogation of the linkages between terrorism, design, and the environment reveals a passage from the unity of a single planet facing an eschatological nuclear winter to a bifurcation of approaches. On one side, the risk society, with a tendency to the extreme, open to developing environmental strategies such as solar radiation modification, welcomes the technological sublime of the nuclear era. On the other, beauty without risk, where agencies turn back and celebrate the era before nuclear weapons, cleanses and neutralizes the wasteland wilderness.

The variety of approaches to managing nuclear waste–from physical isolation to symbolic warning systems–reflects a broader ecological and philosophical challenge in which unity in environmental management is often elusive, and the designed landscape is instead marked by a mosaic of solutions. This unity, which may be understood as clarity for any visitor interpreting a landscape design or an ease with which agencies and stakeholders collaborate on a shared goal for that landscape, is an ideal central to the messages of the 1986 conference and the 1966 and 2016 declarations. Landscape architects could take heed of Karl Linn's intuition to widen the spectrum of voices participating in professional discourse and become a better conduit for that unity. This challenge points to a broader tension between immediate remedial actions and the long-term philosophical and ethical considerations that such monumental legacies necessitate. Ultimately, the legacy of the Manhattan Project and subsequent remedial efforts like FUSRAP reveal a complex narrative of human interaction with time and the environment, where unity, crisis, healing, and responsibility continuously intersect in a pattern of sublime and practical action.

29 *The New Landscape Declaration: A Call to Action for the Twenty-First Century*, xx–xxii.

30 Amy Trachtenberg & Jeffrey Miller, "Leading and Following: Amy Trachtenberg and Jeffrey Miller in Conversation," *The Back Room* (March 12, 2024), https://www.smallpresstraffic.org/the-back-room-article/leading-and-following-amy-trachtenberg-and-jeffrey-miller-in-conversation.

31 Meyer discusses Buell's quote of Ulrich Beck positing "that if only nature is brought into people's everyday images, into the stories they tell, can its beauty and its suffering be seen and focused on." Meyer, "Uncertain Parks," 71.

32 Jason Krupar, "Burying Atomic History: The Mound Builders of Fernald and Weldon Spring," *Public Historian* 29, no. 1 (2007): 31–58.

33 Per communication with DOE Education, Communication, History, and Outreach Team. The University of Cincinnati Center for Design Research and Planning was contracted by DOE to redesign a work on Fernald Preserve. At Weldon Spring, KDG Shive Hattery worked with Stanley Consultants; components of the master plan are credited to SWT Associates.

34 Marj Oaks, "Coming Full Circle to Howell Prairie," *Land and Water* 47, no. 4 (September 25, 2003), https://www.landandwater.com/features/vol47no4/vol47no4_2.html.

BREAKING NEWS
IN CONVERSATION WITH
MICHAEL OPPENHEIMER, ALBERT G. MILBANK PROFESSOR OF GEOSCIENCES AND INTERNATIONAL AFFAIRS AND DIRECTOR OF THE CENTER FOR POLICY RESEARCH ON ENERGY AND ENVIRONMENT AT THE PRINCETON SCHOOL OF PUBLIC AND INTERNATIONAL AFFAIRS.
LA+

Michael Oppenheimer is the Albert G. Milbank Professor of Geosciences and International Affairs and director of the Center for Policy Research on Energy and Environment at the Princeton School of Public and International Affairs. Educated as a chemist and physicist and working at the interface between public policy and science, Oppenheimer has long been a leader in calling for governmental action on climate change. His efforts to bring together scientists and policymakers to elaborate the implications of climate change played a significant role in initiating the negotiations that led to the UN Framework Convention on Climate Change, the parent document of the Kyoto Protocol and the Paris Agreement. **Karen M'Closkey** spoke with Oppenheimer about scientific assessments, uncertainty, and how he has remained optimistic after 40 years of advocacy.

+ References to human-induced climate change from CO2 emissions appeared as early as 1896. However, one of the first to warn about its deleterious effects is a 1965 report on the environment by the US President's Science Advisory Committee. Later, a 1969 memo to President Nixon said we should consider stopping fossil fuel use. When did you become aware of the problem?

I became aware of the issue when I read an article by Gordon McDonald in the alumni magazine *MIT Technology Review* in 1969. His article was about the multiple ways that humans influence, and could influence, the climate globally and otherwise. I am a very calm person. I don't get frightened by big threats. This scared me. At the time, I was in a transition from thinking about science as a career to thinking about science as a tool for public betterment. I received a Ph.D. in chemical physics in 1970 and went to work at the Harvard-Smithsonian Observatory, applying atomic and molecular chemistry and physics to astrophysics. It was intellectually stimulating, but I decided that astrophysics research was costly to governments and secondary in importance to me compared to solving problems burdening people and societies. Thus, by the late 1970s, I had decided that academia wasn't a good fit for me, so after 10 years at Harvard, I left for a job as an activist scientist doing policy-relevant work on acid rain, ozone depletion, and climate change for the next 21 years with the Environmental Defense Fund (EDF).

+ You were one of the earliest and most outspoken scientists warning about climate change. In your congressional testimony in 1986, you said, "If left unchecked [it] will come to dominate all others in its effect on our environment. ...We do not need a comprehensive understanding to undertake action." In 1990, you followed up with *Dead Heat: The Race Against the Greenhouse Effect*, coauthored with Robert H. Boyle. Did you imagine that efforts to reduce greenhouse gases would be so slow given the evidence at that time?

One way that I've managed to work on this problem for more than 40 years is that I didn't give much thought to how long it would take. There was no way to have a good frame of reference for how long it would take the world to recognize a problem like this, to make the institutional arrangements to negotiate with other countries, and to educate people about it. With the encouragement of the foundations that supported my work at EDF, I joined up with several other scientists in 1985 to develop a five-year plan to move the issue from the scientific community to the policy arena. We thought the time for action was ripe because a consensus on the inevitability of climate change and the need for governments to consider action had been reached among scientists. Our objective was called a framework convention, a type of treaty.

It took seven years, but the convention was signed at the Rio Earth Summit in 1992. So, we had the treaty, but very little work had been done politically in the US to get people interested in the proposition. This was the big lacuna that the environmental movement let lie there. People came to me in the mid-'80s and said, "You're trying to generate a treaty on an issue that nobody has heard of, and that would involve tearing apart the fossil fuel industry: you've got to be joking." I guess there was no clear strategy; maybe there couldn't have been one that early. People didn't start waking up to it until we started seeing consequences in the climate that scientists could point to and say, *This is climate change.* So, my timeframe in the early days was probably a decade or two for getting some serious action. Now, almost 40 years later, we have some serious action by governments – though not serious enough.

+ In 1983, the Academy of Natural Sciences report on climate change stated that 3°C to 4°C of warming could have deleterious and inequitable effects globally but did not believe that the evidence supported steps to move away from fossil fuel dependency. The report concludes, "In our calm assessment, we may be overlooking things that should alarm us." Why was the report so tepid in its recommendations?

The thing you have to realize about scientific assessments is that experts' individual biases sometimes influence them, especially when there are only a few experts involved. In that report, which was a mixture of science and policy and economics, there was little understanding or focus on how climate change would affect the average person. Some, like George Woodwell, were pushing to get firmer statements about the aspects of climate change we were certain about scientifically. Certainly, Woodwell believed it was time to start thinking about policy, while others on the panel had more influence over the final wording, particularly scientist Bill Nierenberg, the chair, who was rather politically conservative. And there were two economists, Thomas Schelling and Bill Nordhaus, who were also a little too cautious in their approach. They basically said we need 10 more years of research, which I thought was ridiculous. You don't deal with an earth-shaking problem by waiting: at the very least, you start making plans and laying out policy options while at the same time learning more from vigorous research.

Oddly, but not coincidentally, an EPA report came out either the day before or the day after that Academy report. It was written by a core of experts within the EPA, some of whom were working quietly on both the ozone depletion problem and the climate problem. This was during the Reagan administration, you have to remember. The EPA report had a message that was completely opposite to the Academy report. It said it is too late to avoid significant warming. That was the entry point for me to address the issue in a public way. I used it as an opportunity to write an op-ed for the *New York Times* about these two reports, arguing that there is no point in waiting to take action.

+ The First Assessment Report of the Intergovernmental Panel on Climate Change (IPCC), which provided a strong scientific background for negotiations of the United Nations Framework Convention on Climate Change (UNFCCC), was issued in 1990. In 2023, the IPCC 6th Assessment Report was published. Have the reports changed in terms of how they approach policy recommendations?

The big difference is in the rhetoric. IPCC has become more willing to express statements of fact in language that isn't overly cautious. Its language is not out of proportion to the problem on the various points on which it makes statements, but the statements are clearer and more readily understandable than in the early reports. It is better to do that than to write in a neutral tone or, in some cases, complex and uninterpretable language that the average person, and many experts, disdain.

The part of IPCC that deals with policy is Working Group III. It has expanded significantly. Its authors do a variety of things, like modeling the energy system, projecting future emissions, discussing the positives and negatives of international agreements on climate change, and discussing ways to reach an international agreement. They have gotten more and more detailed, but they are careful to avoid pronouncements about national policy because the IPCC is a collaborative effort between governments and scientists. In other words, the scientists found a client that was interested in receiving the reports. IPCC was put together by the clients: governments. The IPCC reports are the jumping-off point for things like the US national assessment, which provides more detailed, nationally relevant information about climate change. In this way, IPCC has been extremely influential in shaping the body of knowledge that governments look to, and they have taken argumentation about science out of the hands of the negotiators at the conferences of the parties to the framework convention, which is a good thing. Otherwise, a few contrarians in the room would argue about what the science is. Now, the delegates don't have to argue about the science. They take it as a given.

+ You said in your co-authored book *Discerning Experts: The Practices of Scientific Assessment for Environmental Policy* (2019) that consensus is not necessarily the most important feature of assessments. How do you deal with information that does not have much certainty or consensus?

A lot about the climate change problem is very well known within a very narrow range of uncertainty, such as the Earth is warming and it has warmed about 1.2°C in the last 100 years, plus or minus a tenth of a degree; however, you can't reach that kind of consensus with everything. Millions of micro-facts make up the climate problem. For many of these, there is a high degree of uncertainty. We don't have enough observations and governments aren't willing, or can't afford, to pay for dense observation networks. For example, we still can't definitively say whether Southern California will get more or less rainfall because the models can't project with geographic precision, for two reasons. One, the computers are not fast enough, and two, we don't know a lot about the basic science of things you would think we would know a lot about, such as what goes on inside a rain cloud and how a rain cloud forms. So, there is going to be a lot of uncertainty on many specifics. The important thing is not reaching a consensus on every possible statement of fact but reaching a consensus on how big the uncertainties are. IPCC has been doing that rather well. However, even when there is disagreement about the range of uncertainty on a particular factual statement, that disagreement is made public and underscored.

+ Dealing with uncertainty is challenging for designers, particularly when critical factors such as the elevation and datums for building new infrastructure need to be defined.

The issue of how policymakers can grapple with decision-making about climate change is a field that is expanding. There are different models for doing so–mental models as well as computer models–but there is no agreement yet on the best way forward. We're going to learn. We're going to learn by doing in the midst of a growing problem, unfortunately. Our biggest challenge is deciding what actions to take - actions that affect particular people in particular places. For example, our ability to forecast some things, like the height of the surge in Hurricane Francine at a particular place, so you know what the future risk is from other tropical cyclones everywhere along the global coast that are affected by tropical cyclones. In some cases, you need rather detailed knowledge because you don't want to pull people out of where they live unless you have to. Nor do you want to spend trillions of dollars on defensive infrastructure if it turns out that it is unnecessary or should have been put somewhere else. That is where the real battle is today. It's not about whether we should get off fossil fuels: of course we should. It's about how we are going to protect people in this long transition period–both before and after we get off fossil fuels–when the climate will still be warming for decades.

+ We often hear about keeping the global temperature increase below 1.5°C. Where did that number come from?

No scientific law says it must be 1.5°C. There is no threshold to danger - things get continuously worse and the effects catenate. However, the damage increases rapidly when you get somewhere around 1.5°C or 2°C. Some ecosystems, like coral reefs, are already facing severe damage at this temperature. We don't know if significant sections of the ice sheets in Antarctica and Greenland will remain more or less stable at 1.5°C; if they don't, it will lead to continuous acceleration of sea level rise in the coming decades and centuries. In the 1990s, a benchmark of 2°C for the onset of climate "danger" began to percolate among some scientists and policymakers who were looking at things like the potential collapse of the West Antarctic ice sheet. Other changes in the climate system were considered as benchmarks - those that would happen at somewhat higher temperatures and some that could happen at lower temperatures, namely the widespread loss of coral reefs. These impacts would

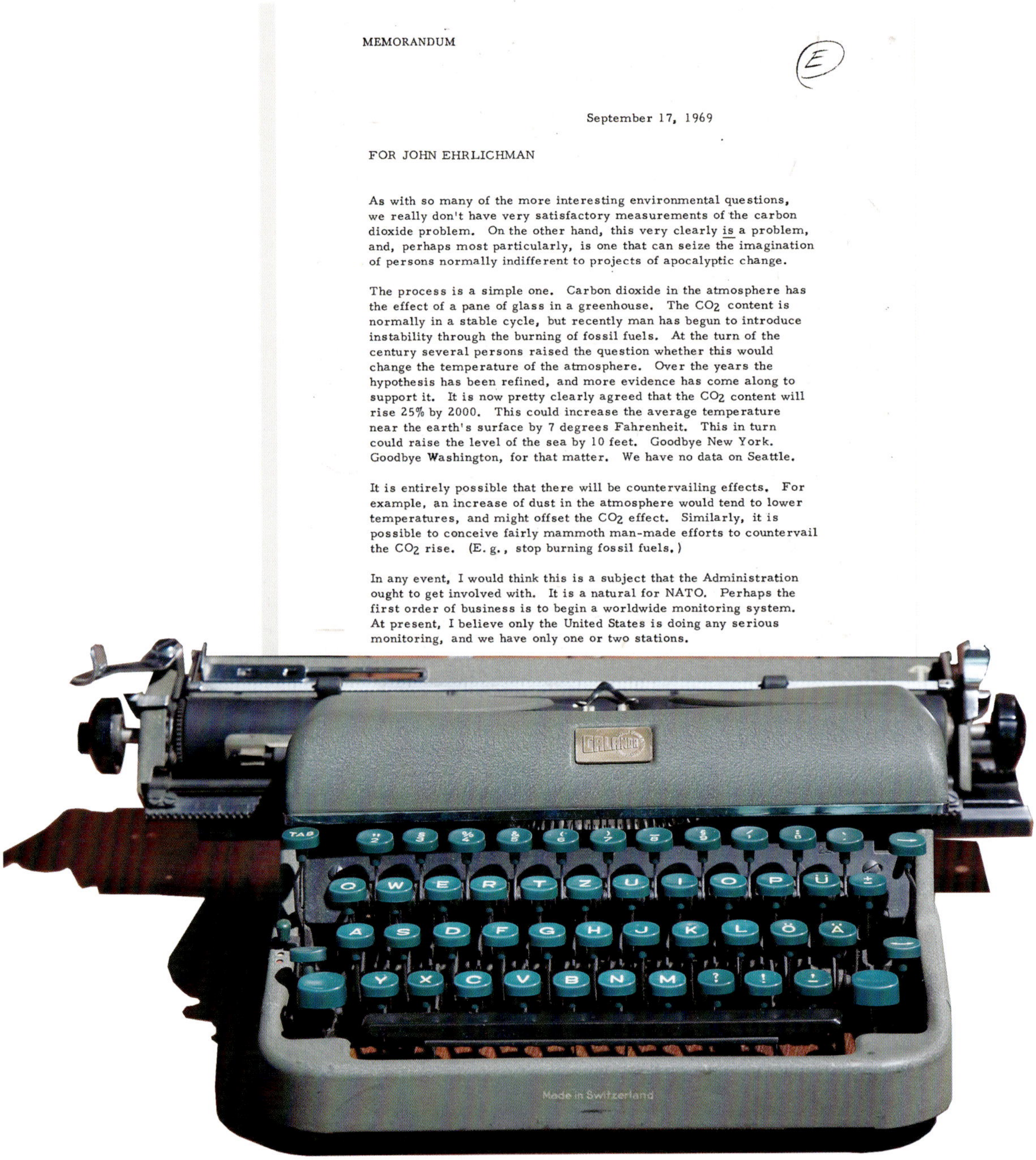

MEMORANDUM

(E)

September 17, 1969

FOR JOHN EHRLICHMAN

As with so many of the more interesting environmental questions, we really don't have very satisfactory measurements of the carbon dioxide problem. On the other hand, this very clearly *is* a problem, and, perhaps most particularly, is one that can seize the imagination of persons normally indifferent to projects of apocalyptic change.

The process is a simple one. Carbon dioxide in the atmosphere has the effect of a pane of glass in a greenhouse. The CO_2 content is normally in a stable cycle, but recently man has begun to introduce instability through the burning of fossil fuels. At the turn of the century several persons raised the question whether this would change the temperature of the atmosphere. Over the years the hypothesis has been refined, and more evidence has come along to support it. It is now pretty clearly agreed that the CO_2 content will rise 25% by 2000. This could increase the average temperature near the earth's surface by 7 degrees Fahrenheit. This in turn could raise the level of the sea by 10 feet. Goodbye New York. Goodbye Washington, for that matter. We have no data on Seattle.

It is entirely possible that there will be countervailing effects. For example, an increase of dust in the atmosphere would tend to lower temperatures, and might offset the CO_2 effect. Similarly, it is possible to conceive fairly mammoth man-made efforts to countervail the CO_2 rise. (E.g., stop burning fossil fuels.)

In any event, I would think this is a subject that the Administration ought to get involved with. It is a natural for NATO. Perhaps the first order of business is to begin a worldwide monitoring system. At present, I believe only the United States is doing any serious monitoring, and we have only one or two stations.

have global reverberations. Then, there was a related discussion of what it would cost to limit warming to 2°C. And a third discussion: Is it feasible to limit warming to this threshold, and to do it quickly? All these considerations went into defining the two-degree target in the Paris Agreement.

But developing countries were annoyed that it wasn't enough, that even at 2°C some of the small island states will lose a lot of territory. If you let it warm 2°C above pre-industrial levels, large areas of Africa will become drier and hotter and won't produce enough food. So, a bargain was struck in the Paris Agreement. The developed countries agreed to set the long-term objective at 1.5°C if the developing countries gave up the idea of deserving financial compensation for the damages that climate change was starting to do. The political reality is that negotiations would likely have collapsed if we hadn't made a significant concession to developing countries – in other words, adopting the 1.5°C target. Many of the rich nations accepted the 1.5°C target pretty much knowing that they would not achieve their fair share of the emissions reductions that were needed to achieve it. Currently, we are on track to exceed 1.5°C, and we are likely to surpass 2°C. With luck and diligent policy implementation, we might end up in the range of 2°C to 2.5°C. But higher temperatures cannot be ruled out; it's possible we could wind up at 3°C of warming, which would portend worldwide disaster.

+ The 1965 report that I referenced earlier suggests that geoengineering–stratospheric aerosol injection [SAI] specifically–might be necessary to counter global warming. You are a chemist and physicist. Should we be studying these technologies?

It's a high bar for me to say we shouldn't be doing research on certain things. The only time I felt strongly we should outlaw research on certain things was, number one, the GMO controversy in the 1970s, when scientists were just starting to think about genetically modifying organisms. I thought the stuff should be banned, at least for the time being. In retrospect, I was wrong in that case because the scientists behaved well and developed a system for monitoring and conducting experiments. I think the world is better off, so far, that genetic modification for medical purposes is now feasible. We might not be having this conversation if the COVID-19 vaccines hadn't gotten out so quickly. Nevertheless, I think a few things are, from many points of view, too dangerous for scientists to research without restrictions, and stratospheric aerosol modification is one of them. At this time, I am utterly opposed to outdoor research experiments because they are a slippery slope from a small-scale experiment that doesn't affect anybody to something that changes climate in a detectable way. I don't trust all the potential players to avoid that slippery slope. Until there can be agreement among scientists, social scientists, policymakers, ethicists, and public health experts–a broad group–over what is safe to do in the open environment, no experiments involving SAI should be done outside.

+ Climate scientist James Hansen argues that we have reached a point where SAI has become necessary. He states that if governments don't take the research seriously, it will not be understood and, therefore, cannot be regulated, leaving tech companies to pave the way.

I'm not sure how much of his view is based on the idea that somebody will do it anyway and it is ungovernable, which I disagree with, or his desperation about how severe the climate problem has gotten. I think he expected people to listen. He expected it to be a rational world, but it turns out not to be a totally rational world. I think he's willing to countenance things on this score that I personally wouldn't. I also believe that barriers to actions by individual bad actors can be constructed. If countries got together and one or two countries wanted to deploy this technology, but all the other countries did not, there is enough power in the group to prevent it. Somebody might do something at a small scale and get away with it, but doing something big enough that it would be possible to detect its effect on the climate, either locally or globally, is very unlikely because, at this point, nobody knows what the pluses and minuses are for specific places, and most countries would not want to bet that they will be among the winners rather than steep losers. Uncertainty is our friend in this case.

+ The UN Environment Programme 2023 report on SAI doesn't rule out small-scale outdoor experiments, but it does say that, once it scales up a little bit, an experiment cannot be differentiated from a deployment. Do you agree?

It is an important point that I and others have made many times. The climate system is so noisy that you could be affecting the climate in ways that are not detectable. By the time you go on for decades, suddenly, you could have pushed the climate into a state you don't want it to be in. I'm basically supportive of more resources to do more and better computer simulations of what a stratospheric aerosol modification, or stratospheric aerosol release, would do. Computer research should push us in the direction of enlightenment about what the effect of a large-scale release is. If, in the process of doing that research, we find that we cannot answer all the key questions we need to know, then we could start discussing what are safe experiments. I firmly believe that something like the Belmont Principles need to be followed if scientists are going to play with the global climate. A few people seem to be willing to put quite a lot of money into geoengineering, some after careful thought and others merely swayed by a belief that silver bullet solutions can save us – which is delusional as far as I'm concerned. It does not apply to the climate problem. Life is more complicated; monolithic, large-scale solutions also tend to entail new and large-scale problems.

+ You mentioned the Paris Agreement. What do you think has been the most important policy or action on climate globally?

The most important thing didn't happen because of the Paris Agreement. It happened sort of inadvertently because China and Germany informally collaborated to decrease the price of solar energy. Germany decided to aggressively subsidize renewable energy, but it became affordable because the Chinese wanted to dominate the international market for mass-producing solar panels and wind turbines. That, to me, is the most important thing that happened to advance the solution to the problem.

With regard to the Paris Agreement itself, the most important thing was that China and the US made a bilateral agreement with specific targets and timetables a year earlier, in 2014. That made countries think that if they went to Paris and negotiated, something good would come of it. Furthermore, at the time, the US finally had an administration in place that was in a position to do something strong due to a 2007 Supreme Court ruling, *Massachusetts v. EPA*. That freed the Obama administration to put a lot of strong greenhouse gas regulations in place using its Clean Air Act authority. The fact that Congress–and later a Supreme Court with a different crew of justices–has become so obstructive to action on climate and environment has limited the amount that can be done using regulation, particularly in the power sector. So, it was very important that the Biden administration, realizing this, decided to throw a huge amount of money after the climate problem under the Inflation Reduction Act. Tax breaks and subsidies without strong regulations accompanying them is a second-best way to solve environmental problems. However, with a second Trump administration looming, it's the most promising alternative we have.

+ Do you mean throwing money at it to change industry?

I mean throwing money and only throwing money isn't enough. What works best is incentivizing research, development, demonstration, and deploying producer and consumer tax breaks and subsidies. However, regulations on what products can be marketed are also needed so industry can anticipate that, as it commercializes climate-friendly technologies like solar panels, heat pumps, or electric cars, consumers won't have the choice to continue buying older, high-emissions products. Getting the timing so that research and development and marketization are synchronized around the exigencies of climate change didn't happen until recently. Now that a second Trump administration is about to take over the government, it will likely proceed to destroy the regulatory structure put in place by Biden.

We don't need to negotiate tougher emissions limits internationally for a while because most countries haven't even put effective and sufficiently comprehensive policies in place based on the commitments they made a few years ago. So, I think it would be a complete waste of time. Instead, the negotiations should be about resolving the loss and damage issue, which is the question of what developed countries will pay the developing countries to help them deal with the problem, even if the developed countries do not explicitly acknowledge that it can be viewed as "compensation," because that word is off the table.

+ As you said, the Supreme Court has ruled on several cases that have significantly limited the regulatory powers of federal agencies for environmental protection, and we are about to have a president who is a climate-change denier. What gives you hope that we can achieve our climate goals?

If we focus on the US, change is well underway. US emissions have been going down for 15 years. Part of that is fortuitous. It was not dictated for climate reasons. It was because of hydrofracking, which replaced a lot of coal with natural gas at a cheaper price. It may, in fact, make the greenhouse situation worse because there are too many natural gas leaks throughout the system, and the leaking gas is primarily methane, a potent greenhouse gas. But that is at least something you can fix. Now the big change underway is stimulated partly by the Inflation Reduction Act (2022) and partly by a deal in the Obama administration, which got the auto companies to agree to clean up their emissions in exchange for the government bailing them out during the financial crisis. Because of that, auto companies have not made a lot of noise about the demand for progressively cleaner cars, which has pushed the industry toward being serious about electrification. Now that China is producing electric vehicles, the auto industry can't go back. There is momentum. There will be stops and starts, like the problem with wind energy and its high capital costs, but overall, renewable energy continues to get cheaper. I think the US is well on its way toward a renewable energy economy in a way that can be slowed but not reversed.

In other countries, there are still issues. India has made a commitment to net zero emissions by 2070, which is fairly aggressive for them, even though it's 45 years in the future. China's emissions may peak in this decade. So, the question is, what is the next country coming over the emissions horizon? We've got to be ready with a better working framework at the international level than what the Paris Agreement offers. There is reason for optimism, if optimism means we'll eventually get the problem under control before it warms 3°C, 4°C, or 5°C. There is no reason to be optimistic about avoiding 1.5°C warming. I can countenance minimal optimism for avoiding 2°C.

RAMI KANAFANI

SPIRITUALIZING THE TWO CULTURES, C. 1978

Rami Kanafani is a PhD candidate in the History and Theory of Architecture program at the University of Pennsylvania. His work examines countercultural architectural practices in the postwar US and their contribution to a spiritually and environmentally defined planetary culture. He seeks to take seriously the spatial and spiritual dimensions of New Age thinking and its focus on ecology in architecture in response to the rise of environmentalism.

+ ARCHITECTURAL HISTORY

In describing the emergence of the environmental humanities in Western academic institutions, Ursula K. Heise draws a distinction between recently formed fields in the humanities–such as disability studies, digital humanities, and human-animal studies, which "propose a new object of study, a new humanistic perspective on a nonhumanistic field, or a particular set of new methods"–and a new environmental humanities discourse that brings issues of climate and the environment to the scale of local communities.[1] For Heise, the field of environmental humanities concerns itself with questions of social and environmental justice, human and nonhuman ethics, and socioeconomic injustices within small communities, thereby departing from its more scientific precedent, environmental studies.

The distinction that Heise foregrounds is the latest commentary on the problem of the "two cultures"–the sciences and the humanities–that British novelist C. P. Snow proposed in 1959.[2] The environmental humanities, for Heise, attempts what Snow thought unimaginable: a synthesis of the sciences and the humanities in the face of an existential environmental threat. The urgency of such a synthesis is the product of the recent awareness, resounding across academic and public life, of anthropogenic climate change and the campaign to codify a new geologic epoch named the Anthropocene. Addressing the imperatives to rethink human/nonhuman/environment relations, the environmental humanities hold as a core principle that issues relating to environment and climate change are, at their origin, of a humanistic, rather than a purely scientific, nature. While possible solutions to environmental issues have been, to a large extent, considered the purview of the methods of positivist science, the environmental humanities undermine the uncritical and ecstatic embrace of institutional, technoscientific solutions in favor of a shift of perspective and scale to individuals and local communities, be they human or nonhuman.

Critiques of more positivist approaches to environmental problems did not originate with the environmental humanities. In the 1970s, members of a countercultural "New Age" group named the Lindisfarne Association (1973–2012) addressed environmental questions by resorting to spirituality, claiming that a revolution in individual spiritual life can usher in a planetary age in which Snow's fragmented two cultures are reconciled. The association was founded in Southampton, New York but soon moved to the Cathedral of St. John the Divine in Manhattan in an attempt to put its ideas into practice by turning the cathedral into a prototypal site of an environmentally integrated community, which they called a "meta-industrial village." They believed that a truly "planetary culture," one that unites the planet's people under the banner of wholeness, could only be achieved by recourse to spiritual traditions that had been excluded from modern life. If, as religion scholar Mircea Eliade claims, "modern man has 'forgotten' religion, but the secret survives, buried in his unconscious," what Lindisfarne imagined was a revolution of consciousness to restore spirituality to a central role in dealing with modern problems, including environmental ones.[3] The absence of serious historiographic considerations of the role of spirituality and religion in modern architecture and in environmental thinking aligns with the tendency to associate modernization with secularization and the diminished influence of religion on life in modernity.

The historiography of modernity–establishing a direct link between processes of industrialization and the rise of new conditions and experiences that are patently modern–has shifted to describe what architectural historian Daniel Barber calls the environmentalization of discourses and disciplines, prime among them architecture.[4]

Again, echoing Heise's point, this does not mean that the environment is a new object of study or a subfield that suddenly comes into the purview of architecture. However, it does instigate a shift of perspective on the history of modernity that accounts for various kinds of environmental thinking, with its multitude of definitions and historical articulations. The environmentalization of architecture history implies that, with retrospective insight from within the Anthropocene, the discipline needs to reckon with its environmental effects and legacies.

One possible avenue is to bring to light neglected or forgotten environmental perspectives on architecture that provide applicable lessons or cautionary tales in the face of an ongoing climate crisis. Lindisfarne's environmental work at the Cathedral of St. John the Divine has been buried with critiques of New Age individualist thinking and thus remains unacknowledged within the postwar cultural reactions to environmental degradation and industrial modernity. On the one hand, Lindisfarne relied on personal solutions to planetary problems and was therefore lumped with other New Age groups for its unwavering faith in individualism as a source of social change. On the other hand, through its environmental and architectural work at the cathedral, it attempted to bridge the gap between the scales of the individual and the planet through local community engagement. Therefore, the rich network of humanists, scientists, and spiritual figures fostered by Lindisfarne at the cathedral warrants consideration for its opposition to the strict distinctions presented by the two cultures and for attempting to bridge local and planetary scales. Lindisfarne's environmental impetus ran parallel to the more political and bureaucratic reactions to the emerging environmental awareness of the 1960s. It acknowledged the ecological limits to unfettered growth under a modern Western paradigm. As such, this episode remains an early example of an attempt to establish cross-disciplinary coalitions in the face of urgent environmental challenges, one that has nevertheless been largely forgotten.

Two Cultures?

Delivering the 1959 Rede Lecture at Cambridge University, the British novelist C. P. Snow diagnosed a "gulf of mutual incomprehension," a proverbial "ocean," between two cultures—the sciences and the humanities—governing Western intellectual society.[5] While these two cultures had managed a frozen smile across this gulf 30 years earlier, "now," he claimed, "the politeness has gone, and they just make faces."[6] At the heart of this increasingly hostile and expanding chasm was a Western educational system that fostered diverging attitudes toward technology and the future. According to Snow, trust needed to be placed in the camp of scientists who, through collaborations with humanists, could restore a global balance for the world's people. Despite admitting the need to bridge the gap between the two cultures, Snow remained ambivalent in his strategies, except for a call to export a Western capitalist and intellectual model to the non-West in an attempt to alleviate global wealth disparities.

Snow's dogmatism drew a barrage of criticism.[7] One of his detractors was the inventor and design scientist Buckminster Fuller, who argued that his own thought and work represented a foil to the framework set up by Snow. Fuller imagined a new kind of subject that he called a "comprehensive anticipatory design scientist" who was well-equipped with the mental capabilities and spiritual dispositions to usher "Spaceship Earth" toward planetary salvation. Despite refusing to describe himself as an architect, Fuller held the discipline of architecture in high regard. He was convinced that "the leaders of the great new technological revolution will come from the architectural profession, which, in an over-specialized age, is almost the only profession trained to put things together and to think comprehensively."[8] In the postwar, he gained popularity with a variety of political and social actors, simultaneously flirting with the world's political leaders and appealing to a dissenting American counterculture who considered him a spiritual father figure who equipped an emerging youth culture with the necessary tools to furnish their alternative lifestyles. His engagements and architectural proposals at the Cathedral of St. John the Divine brought him into the constellation of a class of New Age and countercultural figures rethinking the relationship of the human to the environment and the planet.

An Eiffel Tower in Manhattan

In 1978, Prince Chahram Pahlavi, the nephew of the Shah of Iran, invited a group of 16 scholars from around the world for a conference on his island, D'Arros, in the Seychelles. Sponsored by his Threshold Foundation, the conference's agenda was "to develop a strategy for preserving the earth."[9] Prince Chahram was deliberately vague about the specific details of the conference, imagining that it would "not be a conference at all, but simply an opportunity for a rather innovatory kind of meeting: a 'reverse-prism' for the light of experience of people eminent in their individual fields to re-integrate into wholeness."[10] In true planetary fashion, the issues discussed on this small island in the Indian Ocean had implications on a worldwide scale. On the agenda was the possibility of making solar energy an alternative to fossil fuels, ushering in a solar age after the energy crises of the 1970s. Surprisingly, the attendee list did not include the world's most prominent scientists or economists. Rather, in attendance were a number of esotericists, religion scholars, and theologians such as Mircea Eliade, Titus Burckhardt, and Frithjof Schuon, all from the Traditionalist School, and Reverend James Parks Morton, dean of the Cathedral of St. John the Divine in Manhattan. Buckminster Fuller–arguably the figure most associated with planetary thinking in the postwar era–was also in attendance. It was there that Morton met Fuller and was enthralled by his presence. As Morton recounted in his first of nine "Bucky" epiphanies published in *Buckminster Fuller: Anthology for the New Millennium* (2001), by the end of the conference, he had convinced Fuller to deliver a sermon and to propose a design to complete the unfinished cathedral as a bioshelter ark–a sanctuary for animals and plants–and to update the neo-gothic structure for the solar age in such a way that it would "export solar energy to the surrounding community."[11] In a set of notes taken while on D'Arros, Morton wrote:

> The project goals are twofold: first, to demonstrate through the agency of the cathedral (a national symbol of spiritual/ethical/prophetic values) how a major institution can relate its total program (educational, community, cultural, etc.) to the central issue of the age: the [environmental, crossed out] ecological crisis and alternatives for the future. The point is not to add one more project to an already over-taxed schedule, but to relate all programs to this central focus. ...The second project goal stems from the nature of the cathedral as a religious institution: to bring the ecological crisis of our age into both imagined and religious sensibilities of mankind, and into the central agenda of all organized religious groups. ...The point is to move the world to see the ecological crisis as the religious and spiritual context of our age.[12]

Implicit in Morton's hopes for turning the ecological crisis into the religious and spiritual context of the age is a critique of how possible solutions for this crisis, coming into sharp focus in the 1960s and '70s, had been co-opted by the strategies of positivist science. Morton was calling for what Robert Gottlieb would later identify as an alternative environmentalism, one that would counter a mainstream environmentalism focused on policy and management to the exclusion of other, more humanistic strategies.[13] If this critique sounds familiar, it has the form of the argument, coming some 30 years later, defining the environmental humanities' *raison d'être*: the need to reframe the question of human-induced climate change to the scale of the local community and the individual. For Morton, however, salvation lay in turning to spirituality as a unifying force in the face of environmental disaster.

Fuller proposed a solar tower straddling the cathedral's transept and rising 700 feet into the air. Describing the scheme as a "kind of Eiffel Tower of Stainless Steel," but more modest in size, Morton found that the project achieved "a resacralization of the universe, the rebirth of magic in its best sense."[14] The proposal

1 Ursula K. Heise, "Introduction: Planet, Species, Justice – and the Stories We Tell about Them," in Ursula K. Heise, Jon Christensen & Michelle Niemann (eds), *The Routledge Companion to the Environmental Humanities*, 1st ed. (Routledge, 2017), 1.

2 C. P. Snow, *The Two Cultures and the Scientific Revolution* (1959; repr., Martino Publishing, 2013).

3 Mircea Eliade, "The Sacred and the Modern Artist," in Renata Hejduk & Jim Williamson (eds), *The Religious Imagination in Modern and Contemporary Architecture: A Reader* (Routledge, 2011), 123–24.

4 Daniel Barber, "Environmentalisation and Environmentality: Re-Conceiving the History of 20thc Architecture," *Design Philosophy Papers* 7, no. 3 (2009): 145–60.

5 Snow, *Two Cultures*.

6 Ibid., 19.

7 Literary critic F. R. Leavis unsparingly attacked Snow for regurgitated platitudes and reductive assumptions about humanists and scientists. F. R. Leavis, *Two Cultures? The Significance of C. P. Snow* (1962; repr., Cambridge University Press, 2013).

8 Tomkins, "Profiles – In the Outlaw Area," *New Yorker* (January 8, 1966), 54–59.

9 James Parks Morton, "Nine Epiphanies for Bucky Fuller," in Thomas T. K. Zung (ed.), *Buckminster Fuller: Anthology for the New Millennium* (St. Martin's Griffin, 2001), 78.

10 Letter from James George to James Parks Morton, "Seychelles Retreat Participants," JPM Box #18 File 22, Archives of the Episcopal Diocese of New York, Manhattan, NY.

11 Morton, "Nine Epiphanies," 78.

12 James Parks Morton, Handwritten Notes, JPM Box #18 File 22, Archives of the Episcopal Diocese of New York, Manhattan, NY.

13 Robert Gottlieb, *Forcing the Spring: The Transformation of the American Environmental Movement* (Island Press, 1993), 8.

14 James Parks Morton, "A Symbol Comes to Life," unpublished document, JPM Box #26, Bucky Fuller Folder #4, Archives of the Episcopal Diocese of New York, Manhattan, NY.

15 Ibid.

16 In a letter to Morton, Lindisfarne fellow and New Alchemist John Todd wrote: "I think the Fuller tower project is conceptually poor. It is the old paradigm in new clothes. Fundamentally, it is little different than the Eiffel tower; its message is the triumph of man over nature and God. It even proposes to miniaturize the globe, to shrink the world into a steel and glass ball and pave the way for rocket travel to the heavens. God shrinks too." Letter from John Todd to James Parks Morton (January 18, 1979), JPM Box #2 Folder 4, Archives of the Episcopal Diocese of New York, Manhattan, NY.

17 American philologist and professor David Leeming, an active member of the church, noted that: "To be ecological in the broadest

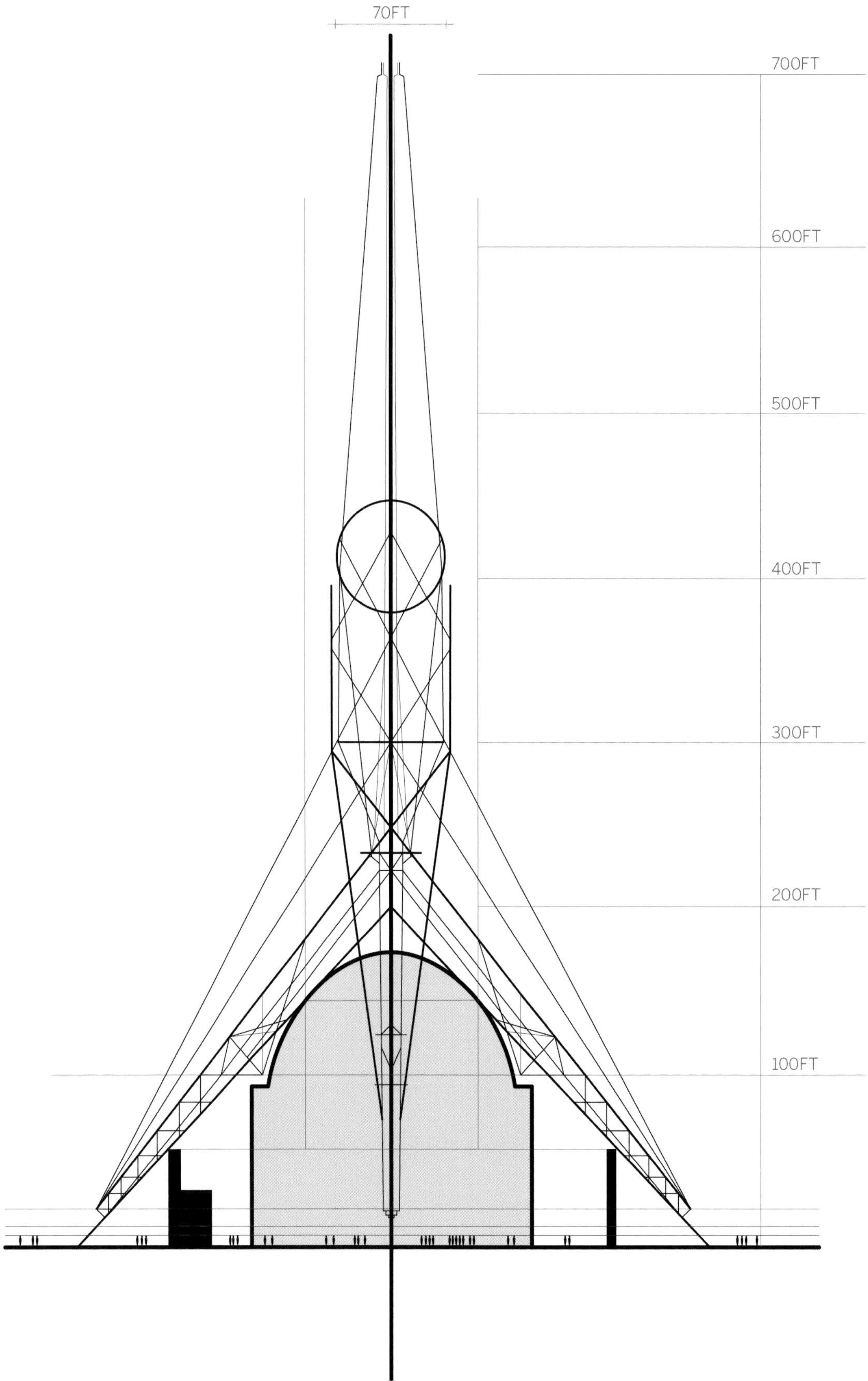
70FT
700FT
600FT
500FT
400FT
300FT
200FT
100FT

was a symbolic gesture pointing from the Earth to the sun, mediating the two realms of the worldly and the divine. Halfway up the tower, Fuller located a Geosphere 70 feet in diameter with a map of the earth projected onto its surface, an iteration of his long-standing Geoscope project begun at Cornell in 1952. The lower half of the Geosphere featured a garden filled with tropical plants. Fuller proposed a bridge running through its diameter where people would stand "at the center of the earth" and gaze at the sky filtered through the map on the surface of the Geosphere.[15] While Fuller's scheme was, according to Morton, symbolically admirable and featured solar collectors on its roof to harness energy, it did not deliver on its promises to export that energy to the community due to the heavy load required for its operation.[16] Seeking to reevaluate the environmental strategies of the church after the criticisms of Fuller's proposal, Morton carved an active role for the Lindisfarne Association that moved its headquarters to the cathedral, where it organized an ecological program with conferences and events that joined scientists and spiritual figures.

The Meta-Industrial Village

The church owed the convergence of ecology and spirituality in the 1970s to Morton and the little-known philosopher and cultural historian who founded Lindisfarne, William Irwin Thompson. They envisioned the church having two functions: the first was to align its mission with pressing environmental questions and to integrate it into its surrounding community; the second, more lofty ambition, was to produce–through the church–a version of environmentalism that they branded "eco-theology," aiming to "mythologize" and "resacralize" modern life through a synthesis of cultural and spiritual traditions that instigate a revolution of individual human consciousness.[17] Thompson was an excellent speaker and proselytizer and was thus able to recruit Morton as a board member for his newly founded Lindisfarne Association in 1973. As a result of becoming a board member, Morton was introduced to the intellectual richness of Lindisfarne's membership, which included scholars from a diverse range of disciplines and backgrounds, including Gregory Bateson, Margaret Mead, Lynn Margulis, James Lovelock, Stewart Brand, Rene Dubos, Fritz Schumacher, Sim Van der Ryn, Paolo Soleri, Francisco Varela, Maurice Strong, David Spangler, and many others.

Lindisfarne grew out of a disillusionment that Thompson felt in a post-World-War-II American society that had lost its way as increasing industrialization gave way to the fragmentation of everyday life. Having graduated from Cornell University in 1966 to become a professor in the humanities at MIT, Thompson lamented the corporatization of education and the alliance of mainstream academic channels to government-sponsored industrial initiatives, referring to the Club of Rome as a paradigmatic example. These elite academic centers, not unlike the urban centers he singled out, such as New York, Paris, and London, had in his mind become hotbeds of industrialization and authoritarian management, far from the integrated communal life necessary to face contemporary social, environmental, and spiritual problems. His critique landed very close to Fuller's as both thinkers grew disenchanted with an American educational system that increasingly tended toward specialization, to the detriment of the kinds of holistic thinking required to face planetary challenges. Not satisfied with simply criticizing the state of things, Thompson "walked out on the university" and proposed the model of the meta-industrial village as a new kind of society. Combining the small-scale economic and social relations theorized by economist Fritz Schumacher for the Global South with the integrated, pre-modern model of the medieval [European] village, Thompson imagined that the meta-industrial village would become a new evolutionary "deme," a term borrowed from evolutionary biology meaning a basic unit of evolution. The deme envisioned by Thompson consisted of the "species plus the environment," echoing prevalent ecological thinking and expanding the focus of evolution beyond the single organism toward a more coevolutionary and environmental paradigm.

Thompson's reversion to the earlier, pre-modern model of the medieval village was not a romantic yearning for an idyllic past. It was a forward-looking projection of synthesizing old forms with new technologies. He explained that history does not proceed linearly and teleologically, nor does it repeat itself. Rather, he saw the progression of historical events as a spiral.[18] His meta-industrial village revived archaic forms and relations and integrated them with the trappings of postwar electronic culture brought about by the cybernetic revolution. He found in the church a connection, at least at a symbolic level, to an archaic past, a pre-modern condition that was not yet adulterated by the fragmentation of modern life. By tapping into past forms and modes of thought and combining them with the tools of information technology, Thompson believed he could recover a lost history and give rise to a new planetary consciousness, one that was centered on the individual human mind first. Only then could it be extrapolated into other nonhuman systems. It mattered little that the technologies celebrated by Lindisfarne were the outcomes of the industrialization that its members so enthusiastically opposed. Such technologies could be harnessed and subverted for positive environmental and spiritual practice.[19]

From Ichthyology to Eco-Theology

Madeleine L'Engle, author and cathedral librarian, described the cathedral grounds as an "island in the island of Manhattan," home to flowering cherry and peach trees, azaleas, daffodils, and tulips and to hens, peacocks, and birds that find refuge in the urban context of Manhattan. But far from isolating the cathedral as an exalted enclave separate from the real world, Morton was adamant about cultivating an inclusive, transdisciplinary culture in which the sciences, the arts, and spirituality converge. In 1979, at the invitation of Morton and Thompson, physiologist James Lovelock, as a Lindisfarne Fellow, presented his Gaia Hypothesis for the first time to the public from the pulpit of the cathedral, anticipating the launch of his popular book *Gaia: A New Look of Life on Earth* that same year. The event was the first in a series of 10 talks by prominent scientists, scholars, and politicians.[20] This was followed up five years later by the formation of the Gaia Institute, which was "dedicated to forging the link between Gaian theory and ecological practice" and worked "in conjunction with public and private agencies" on urban environmental projects. Among them were projects to "rebuild wetlands and revegetate [the] Oakland Ravine in Queens; [to] purify stormwater; [to] design and build food waste compost systems for use on-site in restaurants, schools, and apartment building basements" in addition to a proposal to "make city rooftops ecologically and economically productive."[21] Such active ecological engagements were accompanied by symbolic alliances across species. Starting in 1984 on St. Francis Day, the cathedral held Blessing of the Animals ceremonies where organisms such as elephants, ponies, camels, rabbits, dogs, kittens, goats, geese, hedgehogs, sheep, donkeys, parrots, eagles, fish, tortoises, mosses, blue-green algae, and other varieties of plants were all celebrated in the church. Some organisms found extended residence at the cathedral, like the varieties of fish placed in aquaria within the bays. The directors of the Gaia Institute, Drs. Julie and Paul Mankiewicz, wrote that these aquaria

> [R]epresent a microcosm of the Hudson River ecosystem, from the highest sources of the river to the salt marshes and estuaries that ought to characterize our front door. ...To us, these tanks are no different from the stained glass windows above them. The glass shows the course of Saint Francis's life, his miraculous transformation from a bon vivant wastrel to a saint. The fish tanks show the central and continuing miracles by which our air is generated, our water is supplied, and our earth is renewed. Both of these sequences speak to us of God's activity on the planet.[22]

The parallel drawn between the alchemical transformations of Saint Francis's life and the integrated production of life on earth reflects a synthesis of art, science, and religion into an integrated unity. The "spiritual environmentalism" promoted at the cathedral managed to corral some of the leading thinkers who convened to discuss alternate strategies for dealing with the environmental crisis. These figures moved away from the prevalent technocratic strategies toward more integrated ones: from ichthyology to eco-theology.

Fostering deep associations with prominent New Age figures such as Thompson, Morton worked to amalgamate spirituality with ecological and scientific concerns. According to cultural historian Andrew Ross, the lines of demarcation that science draws around itself are not arbitrary but reflect the play of exclusionary power relations defining the distinctions between science and pseudoscience. New Age culture also circumscribed its own spiritual practices with value-laden lines of demarcation separating it from the worldly and materialist frameworks of science.[23] Based on sociologist Thomas Gieryn's analysis, the boundary work that science—and one can argue New Age culture too—performs is ideological, seeking to construct social boundaries that are not inherent to science [or the New Age] at all.[24] However, New Age culture was strategic in its relationship to a dominant scientific culture; at one and the same time, it appealed to the authority of science and couched its activities in scientific language when it was opportune to do so while distancing itself from rationalist science's more positivist claims on the world.

While New Age culture's moralizing had not produced, according to Ross, "a more consistently accurate account of the world than rationalist science," it did assume "a virtuoso, experimental role in reconstructing a humanistic personality for science – science with a human face."[25] In other words, New Age culture, not unlike what the environmental humanities are attempting to do, humanized science at a time of an emerging awareness of the great environmental threats to planetary systems. Ross had read the New Age as an attempt to carve a place for individualism that, on the one hand, attempted to correct the perceived failures of the collectivist leftist movements of the 1960s, losing hold of personal and private identity and, on the other hand, escaped the New Right's individualist ideology that was subsumed by an urge for neoliberal consumption. The New Age attempted to operate between these two conflicting modalities rising in the 1960s and '80s, respectively, but failed due to its "naïve assumption that changing the self will change the world, an equation that might point to the lack of available languages for linking subjectivity with larger, social or structural change."[26]

At the Cathedral of St. John the Divine, the gap between the individual and the planet was mediated through local community work administered by a spiritually driven New Age collective, one that nevertheless placed the planet's burden on the shoulders of an elite intellectual class. While it is not difficult to recognize the pitfalls involved in the complete faith in individualism as a panacea to social or environmental issues, the critique of the project of modernity that coheres from this episode remains valid. The commonalities between the spiritual environmentalism fostered by Lindisfarne at the cathedral and the environmental humanities lie in acknowledging that the lines of demarcation that science draws around itself are ideological and that addressing environmental problems requires the suspension of such arbitrary distinctions.[27] However, contrary to Lindisfarne's hopes, it might not be in the liberation of the individual human mind that a possible path forward can be carved, given the emerging awareness of anthropogenic climate change and its asymmetric consequences for different groups of humans and nonhumans. As the Anthropocene puts human action in the broader context of environmental and planetary systems, it remains imperative to loosen hardened disciplinary boundaries and to recognize the relevance of the multitude of reactions to environmental problems–from the spiritual inclinations of the New Age to the more recent, non-materialist calls to "make kin, not babies" as a path to ecojustice–while remaining wary of the lofty and hubristic ambitions to "solve" the problems birthed by modernity through spiritual or technological revolutions.[28]

sense is to recognize Holy Creation and our role in it which, given our distinguishing gift of consciousness, must be to make that creation conscious of itself. Religion directs us toward that role by 'raising consciousness' to the ultimate macrocosm, God. To be religious, then, is finally to be ecological, and to be ecological is to be religious. To be religious and ecological is to recognize an all-encompassing pattern in which all parts take on significance only the relation to each other and to the whole." In David Leeming, "Third Series of Notes on a Program of Religious Education for the 'Age of Ecology' at the Cathedral of Saint John the Divine," JPM Box #4 File 20, Archives of the Episcopal Diocese of New York, Manhattan, NY.

18 William Irwin Thompson, *Darkness and Scattered Light* (Anchor Books, 1978), 92–93.

19 William Irwin Thompson, *Earth's Answer: Explorations of Planetary Culture at the Lindisfarne Conferences* (Harper & Row, 1977), 224–25.

20 Attendees included Amory Lovins, Fritz Schumacher, Maurice Strong, Rene Dubos, Hazel Henderson, John Todd, and others.

21 Julie Mankiewicz & Paul Mankiewicz, "Biology in Church," *Cathedral* 9, no. 1 (Summer 1995), 6.

22 Ibid.

23 Andrew Ross, "New Age - A Kinder, Gentler Science?" in *Strange Weather: Culture, Science, and Technology in the Age of Limits* (Verso, 1991), 15–74.

24 Thomas Gieryn, "Boundary-Work and the Demarcation of Science from Non-Science: Strains and Interests in Professional Ideologies of Scientists," *American Sociological Review* 48, no. 6 (1983): 781–95.

25 Ross, "New Age," 29–30.

26 Ibid, 68.

27 The environmental humanities have not entirely omitted religious thought when considering environmental questions. See, for example, Clive Hamilton, who considers the ecomodernist position, which embraces the awareness of humanity's agency as an opportunity to design our way out of the climate crisis, to be a secularized version of Leibniz's Theodicy–the belief that evil in the world is part of an overall benevolent plan. Clive Hamilton, "The Theodicy of the 'Good Anthropocene,'" *Environmental Humanities* 7, no. 1 (2016): 233–38.

28 See Donna Haraway, "Anthropocene, Capitalocene, Plantationocene, Chthulucene: Making Kin," *Environmental Humanities* 6 (2015): 159–65.

Previous: The Fuller scheme of the solar tower at the Cathedral of St. John the Divine.

FAILURE TO BUILD

Holly Jean Buck is an associate professor of environment and sustainability at the University at Buffalo. She is an interdisciplinary social scientist who researches public engagement with climate and energy technologies. She holds a PhD in Development Sociology from Cornell University.

+ SOCIOLOGY, POLICY

Here's where the world is with responding to climate change in the mid-2020s. We had the Paris Agreement in 2015, when countries said they would curb warming to less than 2°C above pre-industrial temperatures and strive for 1.5°C. A few years later, we had over 140 countries announcing net-zero targets – meaning that these countries aim to emit no more CO2 than they can remove from the atmosphere. We had exponentially falling costs of solar energy, with solar becoming the cheapest source of electricity in history. We had youth climate strikes around the world. We had global financial asset managers like BlackRock declare that finance needed to be reshaped to deal with climate change. This all led to some optimism that we were actually going to address climate change.

Then, in 2020, the pandemic hit, followed by Russia's war on Ukraine, destabilizing markets and politics. Compared to the enthusiasm of just a few years ago, the clean energy revolution is stumbling – both in financing projects and in terms of political backlash. Inflation is daunting investors, and many clean energy projects, from offshore wind to hydrogen, have been canceled, largely because the cost of following through with them is far more than expected when plans were drawn up. The rhetoric against Environmental, Social, and Governance (ESG) requirements has led some financial institutions to withdraw from more ambitious pledges and to the pullback of billions of dollars from sustainable funds.[1] Farmer protests have been sweeping through Europe and risk weakening the progress of the European Green Deal. In the US, political uncertainty about the continuation of clean energy tax credits and policies has led to a watch-and-wait mode.

An underlying question at this moment is: Is there the will to rebuild our society? Curbing warming is truly a matter of building and unbuilding. It's about retrofitting or building new factories, generating new types of fuel, reinventing transit, and remaking structures and homes. That's not to erase the importance of institutions and culture that determine the will and possibility to build. Nor is it to minimize the emissions from agriculture and land use change, which are also a third of the challenge. However, a lens that focuses on building can be useful.

Through that lens, progress in the US looks dim. There are many high-profile announcements and engineering and design studies for new infrastructure; however, while some projects successfully make it to a final investment decision, many never pass the concept stage. There's also little public will or imagination of what new factories or clean infrastructure might involve. The *public* is largely disengaged from the project of rebuilding society, except when a project is proposed in a local area. Clean energy projects or industrial decarbonization are often considered part of a different sphere of action, where the responsibility lies with the private sector. Publics may see themselves as "civilians" in the climate fight, to echo a term I've heard people using in focus groups.

It is against this backdrop–where the imagination for building is both privatized and absent–that the idea of solar radiation modification might truly emerge as a solution for responding to climate change. Right now, even though there is a lot of talk, the idea is still highly conceptual. Solar radiation modification, or solar geoengineering, refers to blocking a small fraction of incoming sunlight back into space to cool the planet globally. This could be done by designing a fleet of aircraft capable of flying into the stratosphere and depositing aerosol precursors, most likely sulfates, which are not a novel substance in the stratosphere due to volcanic activity. These aerosols would circle the globe for a year or so before coming back down and, being so high up, can have a cooling effect one hundred times greater than if they were emitted at the earth's surface.[2] This program of constant scheduled flights would need to be maintained until fossil fuels are replaced with clean energy, society increases the capacity of natural and technological systems to remove CO2 from the air, and greenhouse gas concentrations drop. In other words, solar radiation modification involves a multi-decade infrastructure – but one that consists of very little building beyond some new aircraft production lines.

Solar radiation modification thus emerges as a failure to envision remaking the world. It is a failure to finance new infrastructure, to train workers in new practices and technologies, and to develop systems for understanding why we even need this new and rebuilt infrastructure in the first place. It is also a failure in heuristics: people often think about local projects as human-scaled, customizable, and community-produced. Large-scale actions feel imposed. If we persist in this way of thinking about the problem, we will invest in feel-good, boutique-scale climate actions that don't meaningfully make a dent in rising greenhouse gas concentrations. This road leads to despair and proliferating proposals for solar radiation modification.

Solar radiation modification is borne from a scalar hack – a cheap workaround for creating ways of working at the scale of climate action. Yet solar geoengineering faces its own lack of inspiration or will. Right now, few people want to imagine how to do solar geoengineering, and reasonably so. The impulse to consider solar geoengineering comes from a different place than from creative inspiration or imagination. The impulse comes from how hot it is: an emergency drive for shade, for cover, for relief. But it is unlikely that people can build complex multigenerational institutions based upon that impulse, meaning that it is likely that solar geoengineering, too, will be left undone: considered, modeled, toyed with, and abandoned.

This leaves adaptation, which is likely to be the core response to climate change and likely to be unequal and vastly underfunded should trends continue. To be clear, unequal adaptation equals massive suffering. It is not clear that our social systems can adapt to the 3°C or so of warming that we are looking at under current policies.[3] There may be global economic shocks, food price spikes, and an inability to rebuild cities reeling from floods or fires. But to be unable to muster the consensus vision and narrative to rebuild the world to run on clean energy leaves us with this.

If we don't want solar geoengineering or cruel and uneven adaptation, we need to unleash a generative force capable of both the imagination and the follow-through to build at scale. It is difficult to imagine where this energy and vision springs from – "leadership"? "The youth"? Our institutions are not set up for generative work. "Entrepreneurship" and the "dynamism of the private sector" seem better suited for software, cryptocurrency scams, and profit than creative leadership. The government faces many structural, legal, and internal cultural barriers to being a generative force. Much of academia is training people in critique, despite new programs that attempt to be solutions-oriented; it's hard to train people to be visionary when the business model requires large-lecture classes. The incentive structures of our institutions may reward vision *in theory*, even fetishize it, while also being incapable of its real-world application.

"An armature way of building"

Despite these challenges, it is interesting to think about both architects and landscape architects as practitioners in a field that, at its best, translates inspiration into real environments.

Recently, I was fortunate to catch an opening of *Outré West: The American School of Architecture from Oklahoma to California at the Oklahoma Contemporary*, curated by Angela M. Person, Stephanie Z. Pilat, and Marco Piscitelli at the University of Oklahoma. The exhibition focused on a group of architects–Bruce Goff, Herb Greene, Violeta Autumn, Arthur Dyson, Donald MacDonald, John Marsh Davis, and Mickey Muennig–who worked in Oklahoma in the 1950s and 1960s, many of whom later developed design practices in California. They were described as renegades, working within a teaching style that encouraged using organic forms, responding to the natural environment, and developing their own approaches instead of imitating classical or modernist forms.[4] Because the students were driven by their own inspiration, the coherence of the American School wasn't a uniform aesthetic; what held it together was the experimental philosophy and the teaching method centered on student creativity. The school "offered a preview of both the return to the land and the embrace of technology assembled in the Whole Earth Catalog," as noted by Aaron Betsky.[5] That current of fascination with technology interwoven with a sense of place, responsive design, sustainable materials, and experimentation seems to have run dry, either because of capitalism, a technology/nature binary that is still prevalent, teaching styles that foreground critique rather than creativity, or a host of other reasons.

To what extent are ideas in that current worth revisiting now? I was thinking about these lineages one sweltering August morning when I visited the Prairie House, an iconic cedar-shingled home in Norman, Oklahoma. It was designed in 1961 by Herb Greene for his family and looks like an organic prairie creature, maybe a cow or a chicken, depending on whom you ask. Inside was a 1981 volume by Herb Greene with Nanine Hilliard Greene, *Building to Last: Architecture as Ongoing Art*. In it, the authors describe a vision of "an armature way of building," in which an "armature" is defined as "a public element in a neighborhood or city core to which space-enclosing structures and ornamental surfaces of individual determination can be added or subtracted."[6] Key characteristics are that it is long-lasting and "richly encrusted with the crafts and arts of as many as thousands of participating citizens."[7] The armature is imagined as an accretion of citizen art that will take generations or centuries to build.

Greene saw how building without the public was an issue, but was trying to articulate a way forward, especially regarding megaprojects: "Armatures are an opportunity to make a synthesis of three technologies: our present high technology, an intermediate technology suited to small groups of citizen-builders, and the return of handicraft. The goal is to bring into the building system thousands of people whose talents are now excluded."[8] To further the inclusion of citizens in building, new *social* technologies would be needed. Greene refers to new design processes like town halls and participatory workshops that the "architect-coordinator" would utilize. Note this new role doesn't really have a name, even today – it is much more involved than "community outreach."

Even back in 1981, the vision seemed utopian, leading Greene to admit in the preface, "The more I try to hypothesize how on-going structures, which I call armatures, can actually be built, the more I realize the difficulties that tax the credibility of the theory."[9] But he also points out that many of the strategies and institutions that are prerequisite to them exist. Today, the renewed interest in handicrafts could also facilitate the development of armatures. Yet, it is even harder to write utopian ideas in a way that would be credible. We may already be too constrained to think and vision our way out of climate change.

Decarbonization and scalar fluency

Nearly half a century later, where are we? Two challenges loom: public participation and working at a planetary scale. At first glance, they are directly in tension with each other. When it comes to the challenge of creating a public sensibility around inspiration and building, there is still obvious interest in what we might call sustainable participatory design, and no shortage of writing and practice. But we are still mainly failing at it compared to the immensity of the challenge.

Dealing with scale is the second challenge, and it involves both building at scale *and* thinking at scale; that is, letting scale drive decisions beyond the individual project. For example, there is a lot of excitement about mass timber, including its potential as a carbon-removal solution. In one way, this is wonderful. But there is a scalar challenge, and it points to a need for bigger-picture planning. The new terminal in Portland Airport was unveiled in 2024 – a compelling, airy lattice of light-colored wood. The headline in the *Seattle Times* read, "It's stronger than steel, lighter than concrete and captures carbon – mass timber is the future."[10] Architect and proponent of low-carbon ecodistricts Mike Eliason weighed in with some observations on the total carbon effects of the new airport terminal – it may sequester some 2,300 tons of CO2, but that embodied carbon is a minuscule fraction of the airport's impacts, given that most of the emissions are due to the planes taking off.[11] It may echo Greene's 1981 warning: "It is also possible that we may continue to build irresponsible, energy-wasteful structures and camouflage them under superficially 'ecological' shapes and surfaces." It is not enough to have beautiful, inspired designs of buildings that sequester carbon – there also needs to be a scalar sensibility, a scalar expertise, that informs what to do at the global and national scales, and to connect with landscape-level decision-making. Obviously, with the division of labor in society, those kinds of decisions are outside the purview of the architect, who is hired later in the process. We don't even have the planning capacity to guide national decisions in line with what scientists say would be needed to curb emissions to safe levels. On the other hand, some of the speculative projects and design competitions *do* attempt to address scalar thinking, such as SCAPE, the landscape architecture and urban design practice of Kate Orff, the McHarg Center's *The 2100 Project: An Atlas for the Green New Deal*, or the Berggruen Institute– all of which approach the scale aspect of ecological challenges in different ways.[12]

Both climate change and the energy transition demand thinking about scale in new ways. The electric grid. Landscapes of fire and fuels management. The foodshed. With carbon constraints come scalar constraints because the life cycle analysis of the carbon becomes primary and impacts where materials can be sourced. This is relevant for things like bioenergy, renewable natural gas, and hydrogen, and is obvious when it comes to carbon removal.

To reach net zero greenhouse gas emissions by 2050, which is what the US is aiming for (along with many other countries), some amount of remaining emissions from sectors that are difficult to decarbonize would be balanced by removing carbon from the atmosphere. Countries that have set net-zero targets implicitly assume some amount of carbon removal capacity. Carbon can be removed using biological approaches, like planting new trees or farming in ways that put more carbon back into the soil, though these approaches generally store carbon temporarily. It can also be removed using engineered approaches where CO2 is injected into rock formations deep underground, where it eventually mineralizes. For example, biomass could be harvested and combusted in a power plant, where the CO2 could be separated out, transported, and sequestered geologically. But for this to be a carbon-negative process, there couldn't be too many emissions from harvesting or transporting the biomass – those almond shells, pine trees, or other feedstock would need to be local and transported using clean energy. Or, if one wanted to build machines that capture carbon directly from the atmosphere, those machines would need to run on renewable energy for the process to remove maximum carbon. Soon, the field of building is not a biomass power plant or a direct air capture unit but a solar farm, an almond orchard, a processing plant, and more. The scale can be expanded, but our institutions are still oriented toward the project level.

Indeed, our institutions of building, including finance and policy, are not working through these linkages across scales. Another example is the US government's investment in hydrogen. Part of the attempted strategy to deal with the need to fund companies that source hydrogen and find companies that will buy hydrogen is to create "hubs." Hubs are a strategy with direct air capture, too. But no one really knows the scale of the hub – some of the hydrogen hubs so far involve multiple states. We are grasping for a scalar sensibility, for a new way of seeing how the pieces fit.

But the tension remains. When we think about scalability in a time of urgency, we tend to think about the efficiencies that come from mass production and replicable designs. That seems to be in direct conflict with the bespoke process of community-engaged building, with the idea of an armature that everyone contributes to. Can scalar fluency and citizen participation work together? What if participation is not an obstacle to dealing with the challenges of scale but the unlock for scalability?

Right now, the dominant tendency in building is to minimize frictions with the public, and treat them as obstacles to brush past, continuing the "decide-announce-defend" model. However, the public's potential demand for the new may actually be an untapped resource. This is the basic theory of the Green New Deal – voter desire and support leads to the public spending needed to catalyze building at scale. The fact that in the US, ambitious Green New Deal-type legislation was discussed at the outset of the Biden presidency and shrunk down into the Bipartisan Infrastructure Law and Inflation Reduction Act does not mean that the theory is wrong. There are at least two elements to making participation cohere with scaling. One is the education needed to understand the scale of the problem and allow that to be design criteria. Even at the university level, this is not well taught, and the quantitative reasoning that informs scalar intuition needs to be taught in primary and secondary school.

The other element is to have a catalog of basic building blocks that can be used and shaped by communities in meaningful ways. Let us take two very different examples of this. Consider a biochar pyrolyzer, such as the Pyreg PX-1500, manufactured by a German company, which produces biochar from biomass. It also produces thermal heat that can be used in district heating, and this unit serves the average demand of around 180 households while removing about 2,000 tons of CO2 per year. It is modular with plug-and-play installation.

Next, consider the AP1000 nuclear reactor, the Westinghouse design powering the new units at the Vogtle generating plant in Georgia, which came online in 2023 and 2024, and produces 1.1 gigawatts of power, which is enough for 500,000 to 750,000 homes. The US will need at least 700–900 GW of additional clean firm capacity to reach net-zero, according to the Department of Energy analysis, and nuclear power is one option that can achieve this scale.[13] Some analysts would say that despite the enthusiasm for small modular reactors in tech circles, this larger design is what is realistic to meet the goal of tripling nuclear capacity by 2050 – a goal that the US signed onto along with over 20 other countries. Including nuclear in a system with renewables and energy storage reduces overall system costs by 37%, which is a major consideration given public concern about the cost of the transition and who might bear it.[14] So, there is a case for pursuing large-scale nuclear projects.

The AP1000 is never going to be an armature, for security and regulatory reasons. Yet even though it can't be physically customized and built by a community, community engagement and input are critical to getting these units built. This is the lesson learned in places like Sweden and Finland, which innovated processes for siting that involved compensating communities for participating in feasibility studies and environmental reviews and giving communities veto power.[15] The US Department of Energy is currently funding a consent-based siting approach to community engagement around nuclear waste, figuring out the need for input throughout the project cycle and beyond the project site. The point is that even when imagining a large-scale project with a replicable model, public engagement is critical *around* the physical unit – thinking about the benefits, the businesses nearby, the social relations, and the landscape of where the project is situated. New AP1000s can provide clean baseload power that will power other social aims, and the incomes and revenue from operating new plants could bolster budgets for new parks or community projects.

The Pyreg PX 1500, on the other hand, could become part of an armature, a public element within a connected whole. It can fit in a relatively small building. One can imagine it as part of an integrated forest campus that makes other sorts of products, a forest or biomass worker training center, a place where gardeners can pick up biochar, an educational facility, or more. It could be a building block that the community gradually adds to over decades and become part of the fabric of community life – if that were how we chose to build. The blueprints for these basic units exist. The question is whether we can create institutions for collectively imagining the rest of the social and ecological landscape and consider how large-scale action works with community involvement. An approach to the energy transition that opens up an armature way of building may get us out of solar radiation modification and uneven adaptation.

1 Economist Intelligence Unit, "Anti-ESG Sentiment in the US Weakens ESG Markets," https://www.eiu.com/n/anti-esg-sentiment-in-the-us-weakens-esg-markets/ (accessed Sept. 19, 2024).

2 David W. Keith & Wake Smith, "Solar Geoengineering Could Start Soon If It Starts Small," *MIT Technology Review* (February 5, 2024), https://www.technologyreview.com/2024/02/05/1087587/solar-geoengineering-could-start-soon-if-it-starts-small/.

3 *UNEP Emissions Gap Report 2023*, UN Environment Programme, https://www.unep.org/resources/emissions-gap-report-2023.

4 Stephanie Pilat, Angela Person & Marco Piscitelli, *Outré West: The American School of Architecture from Oklahoma to California* (Oklahoma Contemporary Arts Center, 2024).

5 Aaron Betsky, "Foreword: Questions Concerning the American School," in Luca Guido, Stephanie Z. Pilat & Angela Person (eds.), *Renegades: Bruce Goff and the American School of Architecture* (University of Oklahoma Press, 2020), xi.

6 Herb Greene with Nanine Hilliard Greene, *Building to Last: Architecture as Ongoing Art* (Architectural Book Publishing Company, 1981), 9.

7 Ibid., 9.

8 Ibid., 14.

9 Ibid., 5.

10 Josh Farley, "It's Stronger Than Steel, Lighter Than Concrete and Captures Carbon – Mass Timber Is The Future," *Seattle Times* (August 9, 2024), https://www.seattletimes.com/opinion/its-stronger-than-steel-lighter-than-concrete-and-captures-carbon-mass-timber-is-the-future/.

11 Mike Eliason on X, "Let's talk about the total effects of Portland's new mass timber airport terminal," https://x.com/holz_bau/status/1824121187448107472 (accessed Sept. 19, 2024).

12 See SCAPE, https://www.scapestudio.com/; The McHarg Center, "The 2100 Project: An Atlas for a Green New Deal," https://mcharg.upenn.edu/2100-project-atlas-green-new-deal; Berggruen Institute, https://berggruen.org/about.

13 US Department of Energy, Pathways to Commercial Liftoff: Advanced Nuclear, Sept. 2024. "Clean firm" power refers to zero-carbon energy available whenever needed.

14 Ibid.

15 Jessica Lovering & Judi Greenwald, "Efforts to Accelerate Permitting Could Learn a Thing or Two From Nuclear Energy," UtilityDive, https://www.utilitydive.com/news/permitting-transmission-renewables-nuclear-energy-advanced-reactors-nuscale-terrapower/648358/ (accessed Oct. 22, 2024).

IN CONVERSATION WITH **DIPESH CHAKRABARTY**

Dipesh Chakrabarty is the Lawrence A. Kimpton Distinguished Service Professor in History at the University of Chicago. A renowned scholar, he is the author of numerous essays, books, and edited volumes. His most recent publications include *The Climate of History in a Planetary Age* (2021) and *One Planet, Many Worlds: The Climate Parallax* (2023). These works challenge scholars to rethink human history in the context of rapid climate change. **Karen M'Closkey** spoke with Chakrabarty about how his work has evolved, and the philosophical and political problems that arise when historicizing within a planetary context.

+ Your work over the past two decades has centered on how anthropogenic climate change disrupts the previously accepted distinction between natural history and human history. Before this, you gained recognition for your contributions to subaltern and postcolonial studies. Do you see your later work as a departure from this early work?

It was a departure in that when I was working on subaltern studies and provincializing Europe—looking at normative categories of modern European thought as a necessary but insufficient resource for understanding lives and institutions in India and elsewhere, including in the West—I was focused on differences and inequalities between humans. Though I was aware of environmental histories, I was not an environmental historian. Coming from the Marxist left, I thought that the main problem was capitalism and that fixing capitalism would at least create the opportunity for fixing other problems. Capitalism is still important to study but it is embedded in the Earth system or what I later called the planet – the geobiological processes that support life in general. This planet was a taken-for-granted entity, though we did not think about it consciously in that way. My later work on climate change allows for the role consumerist capitalism played in creating the climate crisis, but it is also an exercise in provincializing humanity. I look at the human story; it is a form of life that has always been a minority form of life that hasn't played a great role in keeping life going on the planet but has now expanded its realm so much—thanks to industrialization—that it is threatening the prospect of life on the planet. What I try to retrieve out of my present preoccupation is a reminder that we are a minority form of life. In many ways, bacteria, viruses, phytoplankton, and other things we consider to be inferior forms of life have historically had a much more important role in creating conditions for life's flourishing. So, in that sense, I used to provincialize Europe, and I think I provincialize humans now in that I find anthropocentric categories necessary but insufficient in understanding the full significance of our planetary environmental crises.

+ You describe "the planetary" as an emergent category of humanist thought and have argued that it is distinct from "world," "earth," and "globe." What are some key differences among these terms?

What I call the globe is made possible by technological connectivity. Everywhere has a human footprint because of empire building, slave taking, colonizing, and appropriating Indigenous people's land, but also building capitalism and trade networks. This technosphere is the globe. Technology is an infrastructure we built to extend human life, but it sits on another infrastructure that nature has built, which maintains life in general. This Earth system is much bigger than the globe. There is a conflict between the two infrastructures because, for the past two hundred years, we have built our infrastructure by being oblivious to the second, more important infrastructure underneath, which is the planet, marked by a deep interaction between bio-geological forces. The globe and the planet give very different perspectives on human beings. Of the globe, we are the protagonists. We created it. Of the planet, we are beneficiaries. So, forms of life we consider inferior keep supplying the air with fresh oxygen as they have done for four hundred million years. They haven't done it with us in mind.

+ The meaning of the word "environment" has evolved significantly. Etienne Benson's opening essay explores this transformation, one of which is that, with the rise of Earth system science, the environment has taken on a global scope. The concept of the global environment sounds a lot like "the planetary." Are they synonymous?

They are related but not synonymous. I think of the planet as a scalar entity. In a sense it is very large, the whole Earth taken as a system [by climate scientists] or the Earth system. But it exists on the scale of the personal too. Your microbiome is planetary; it is not a global environment. It is your internal environment, if one may say so. And there are regional environments that have planetary significance but are not global in scale – say, the environment constituted by Himalayan glaciers. Secondly, the planetary is not always indexed to human experience. The deep oceans have a role in maintaining the climate system, but we do not experience the deep ocean. So, the planetary involves the many processes of which we are beneficiaries but that treated us as marginal to their own operations until we became technologically powerful enough to break them down at some point.

+ You have said that while climate change is not a shared responsibility–given that wealthy nations are liable for creating this crisis–it is a shared predicament, and thus, a critique of capital is "not sufficient for addressing questions relating to human history." Some scholars have criticized this viewpoint. What is your response?

The key difference, as I understand it, is this: my critics' argument is that human flourishing based on inequality has been the primary cause of the environmental crisis. The assumption is that if humans were more egalitarian and more oriented to social justice, the planet would not be in this crisis. There are two problems with this. One is that even a relatively egalitarian human world could produce more greenhouse gases if humans decided to consume a lot of fossil fuel for their own needs of flourishing. But the second and more important problem is that even if we could avoid overconsumption and seek alternatives to capitalism–the critique of capitalism is not wrong–humans would need to be collectively rational in globally regulating their lives. There is no historical evidence that they are. For example, Trump walking out of all climate deals and actively undermining them is an instance of tremendous irrationality that has political clout. We do not have societies based on collective rationality. How do we go from a society that is not based on collective rationality to one that is? Revolution is one answer. The idea was that, by force, we would make a revolution and then impose collective rationality, but who would guarantee that the regulatory institutions would not be gamed? The climate question confronts us with the reality that even if the planet were to be considered a single climate system, there is no corresponding single humanity to deal with it. This is a major philosophical, anthropological, and political problem.

+ It is interesting how much of your work draws on ancient and Enlightenment thinkers more than on ecophilosophers or environmental historians. Given that the "human condition has changed," as you say, how do their ideas still hold power?

Well, I think their ideas do hold power because–and this is where my work is different from eco-philosophical work–Enlightenment philosophers were asking a lot of meta-questions about the history of modernity. The meta-questions are about freedom, autonomy, and respecting human beings, not as a means for something but as ends in themselves. Those values were very important to my work in postcolonial thinking. Regarding the climate work, what fascinates me is human phenomenology the human experience of the world limited by our perceptual apparatus–and the relationship between those limitations and what we cognitively learn. Cognitively, we work out that the planet's climate acts like a whole - a "climate system." We don't experience the climate system. Similarly, we speak of an Earth system where the geological and biological processes come together. These are what we know cognitively, not what we experience. Instead, we experience seasons or beautiful or bad days or, now, erratic weather.

These perceptual limitations are central to how we experience the world and how we have evolved. Ingrained habits have to do with evolution because they must be habits that have helped us survive. At the same time, we are now faced with a problem that challenges the habits or instincts that have evolved over evolutionary timescales. So, what I bring to bear on my thinking is this whole question of deep history, which is much deeper than the history of capitalism and the history of contemporary inequalities, but which runs through capitalism, through contemporary inequalities. And again, to go back to my early work, it gave primacy to the institutions of capitalism, thinking that getting rid of capitalist institutions would solve our problems. Now, I see this crisis as challenging our evolved habits of thought and action and raising the more philosophical question of what it means to be human. I don't think what I am saying is at loggerheads with what my critics are saying, but sometimes the argument is at cross purposes. The questions that interest me are not necessarily the questions that are of interest to scholars who are more keen on finding in capitalism an answer to the question as to why we're in this situation and potentially a way of solving the problem.

+ You have written about how the rise of secular thought and becoming modern has diminished fear, leading to a loss of reverence for the natural world. Our efforts to control nature have contributed to our current climate crisis. Can fear inspire a new reverence that avoids seeking large-scale global solutions?

If you go back to the world, let's say before industrialization, and look at peasant societies or Indigenous societies, you will find that the seasons were of great importance to agriculturalists and people who went trading by sailing boats. You were dependent on when the trade winds came and how they blew. You were dependent on the currents in the ocean. And so, I would say that most pre-industrial societies were aware of what I call "planetarity" because they depended on the planet doing its work. And in those societies, you see one way in which they kept the human realm delimited and small because they didn't have the technology to expand it was through injunctions - don't do this, don't do that, because that could cause trouble. And those

injunctions sometimes take the form of what an Enlightenment person might regard as superstition. But one function of superstitions was to keep the human world delimited. These societies supported smaller populations, and fear was one way to keep the human realm safer. The Enlightenment was described as the overcoming of this fear. Some of the overcoming of fear was good for people. For instance, the history of modern medicine is full of fights against ideas that modern medicine found wrongheaded. For colonized people, overcoming the fear of colonizers was very important politically. On the other hand, human fears of other creatures kept them safe from humans, too. These kinds of checks and balances were lost when we developed technology that made us, collectively, a threat to wildlife. You can look at the history of urbanization and industrialization as overcoming our fear of other species and eventually overcoming our fear of the planet. The idea that you can travel by train or fly in any weather was celebrated as freedom. Before that, travel was seasonal. In India, before the British came, even wars were seasonal. So, technology is a history of forgetting the planet. But now, through this crisis, technology has reminded us that the planet exists. We cannot go back to the fears of the old kind because of what we know now. But I often think that the secular form that fear has taken in modern society is the idea of risk.

+ Risk is based on the rise of statistics and probability. It is another way to tame things by producing averages and norms around which we build our infrastructure. As you say, the challenge is that we cannot go back, technologically speaking, but we are in this perpetual cycle of scaling up.

I agree. Modern societies are based on risk management. Life insurance, pension funds, home insurance. Managing these risks is easier to do when the planet is not a bigger risk to you. But if you have warmed up the planet, you must take even bigger risks to keep the small risk management going. Then you end up with geoengineering the climate of the whole planet so that we can continue our lifestyles. We end up wanting to manage even bigger risks in order to make life seem normal.

Recently, I have been looking at Bill Gates's *How to Avoid a Climate Disaster* (2021). He is a smart, good-hearted philanthropist, but he is also devoted to capitalism and techno-fixing. At the same time, I've been looking at Marxist de-growth literature. These different worldviews would agree on some things. Nobody wants infant mortality to go up. Nobody wants MRI machines to go away. The fact that you can now add several years to a person who is 85 is, socially, a very expensive proposition, but we, as biological creatures, love those extra years. Capitalism and technology have reached a point where they speak to certain fundamental biological drives within us. If birds could invent this technology, I think they would do it. We happen to be the only creatures that have succeeded in creating longer lives for ourselves, and that is irrespective of whether you believe in socialism or capitalism. Justice requires that everybody should be able to live the full term of their natural life as made available by technology. It is a demand made by Amartya Sen, Martha Nussbaum, and human development scholars that longer and healthier lives should be available to everybody. However, longer and healthier lives are owed to technology and economic growth.

That is the depth of the predicament that we face. As a historian and philosopher of history, this renews the question of what it means to be human. I am not a technical person. I am not a policy person. I love thinking about history; what are the stories that humans tell themselves about themselves? I think we have reached a point where that story must be renarrated and retold. And one of those stories is about how divided our politics are and how much politics itself is based on divisions, in good and bad ways. A democratic dialogue is based on the capacity to disagree, however, the disagreement of a very powerful person carries more weight than the agreement of less powerful people. That is where things get scuttled, which sometimes makes me feel that we may not be able to deal with the Earth system in time.

+ We have seen how efforts to address serious global issues, such as the Green Revolution to combat world hunger and recent forest management practices focused on carbon accounting, have led to severe social and environmental consequences. Does planetary thinking risk "reducing the future to climate" to use Mike Hulme's phrase because it subordinates non-climate issues and narrows political possibilities?

It will reduce the future to climate if you think of climate in isolation from other things. But if you think of climate as something produced by life, that life has a role, like the phytoplankton that produce the fresh oxygen supplied to the air every day, then climate is not an object but an entry point into the Earth system. The techno-fixers sometimes fetishize climate. In fact, they don't fetishize climate. What they actually fetishize is temperature. Those who promote geoengineering think the job is to cool the planet down by dimming solar radiation. They think temperature is the most critical variable; bring it down, and then you have time to deal with other problems. But they assume that the planet would be just sitting quietly in the meanwhile. We have no idea what the effect would be on other life forms if we do this. Knowing about the importance of biodiversity, both in the history of life and in human history, should be compulsory for all disciplines and education. We should teach children about biodiversity in the way that we teach them numbers. You cannot operate in modern society without numeracy, and we are moving toward a point where we won't survive if we don't acknowledge the importance of biodiversity. For example, at the University of Chicago, we have set up a climate energy and sustainable growth institute, and its pillars are economics and policy, battery technology, and geoengineering. Those are the three verticals. They are important. But the planners of our university don't yet seem to realize how important biodiversity is – it should span across these pillars. And Earth science, too. Science itself has returned to a kind of holism. Philosophically, I see no difference between what Donna Haraway and Bruno Latour are saying and Earth system science, developed by NASA in the 1980s as an interdisciplinary science, says. Earth system science is constitutionally multidisciplinary, holistic, and, therefore, open to the social sciences and humanities. So, in some ways, the techno-fixers are still thinking through older disciplinary cleavages. We cannot eliminate disciplines altogether, but we should build more corridors between them.

+ That is the aim of *LA+*! In this issue, Laura J. Martin argues that E. O. Wilson's "half-Earth" proposal maintains an untenable nature/culture divide that treats *habitat* as a noun–a contained space–rather than think of *inhabiting* as a verb, which can accommodate many ways of being in relation to nonhuman life. Do you think Wilson's framing is a step toward mutuality, or do you agree it is an outmoded way of thinking?

I think the criticism is right, but I have something to say in favor of Wilson. It is a good philosophical message about human withdrawal. Half-Earth metaphorically means we don't need the whole Earth. I think that message was correct, particularly when things like geoengineering and people like Elon Musk are inviting humans to expand our realm to asteroids, the moon, and other planets. Wilson was saying, "No, we need to shrink. There has been an overshoot of the human species." But the critique has a point too, when considering the title of Wilson's book. It remained rhetorical. His practical proposals–such as the one about returning all National Parks to pristine environments–fell short of what the ambitious title suggested.

+ Yes, it is a question of how conservation takes that message of half-Earth and deals with it on the ground, and that brings me to your distinction between the *possible* and the *feasible*. You've said that critiques of capitalist globalization are needed but are inadequate in the face of rapid climate change.

Yes, the important question is, "How do we get there?" I take the idea of path dependence from economists. In other words, your initial conditions determine, to a large degree, how you get from A to B. It is easier to answer the question of path dependency and degrees of freedom at local or regional levels, levels below the planetary. The planet exists in your body; it exists in everyday life in many different forms. Some places may get a messianic leader like Gandhi, who might be able to walk them out of the trap in which we find ourselves, but we are too many in numbers to have one messiah. The messiahs will fight amongst themselves about who the real messiah is, and there is also the risk of messianic leadership being authoritarian, so that is not a solution to the problem. However, if you ask what is planetary about a local problem and address that, I think we can do a lot. If we cannot synchronize human actions at a planetary level, the big problems will continue, unfortunately.

+ For the final question, I would like to return to the first question about what prompted the shift in your work. You have said your awareness of climate change science in the early 2000s and your personal experience with bushfires in Australia at the same time instilled in you a feeling akin to falling. Could you elaborate on this?

Falling is the expression I use, which has a certain affiliation to Heidegger's work. He talks about how being born is this experience of "thrownness" – of coming out into the world whether you want to or not. Here I was, feeling completely comfortable with my work on equality and justice, which gave me a sense of compass in navigating the world without realizing that I had taken the planet for granted. And to realize that the planet didn't really care much about these human questions and to see how fast climate change was happening was a sensation of thrownness, of falling, being thrown into an awareness of deep history and how that history mattered. Whether climate change belongs to the future or the present is a function of how privileged you are. There are places in Rajasthan where the temperatures in summer go up to 54 degrees Celsius. This is not survivable, even with air conditioning. Women in the Bengal Delta still catch small fish with baskets, standing waist-deep in the water. The water has become salty, gets into their reproductive organs, and creates all kinds of infections. The planet is personal. That is why we need a new kind of anthropology, a philosophical anthropology of human beings. Hegel always thought that the cunning of reason would save us in the end, for it was God who endowed humans with reason. A Hegelian might say, "We are entering holistic thinking through our science and technology. We are undoing the forgetting of the last 150 years because reason has reminded us of the planet through this huge crisis. We will be all right in the end." We have become aware of the crisis through disciplines that were created in the 19th century. We created more and more specializations and experts. But I think the fact that the scientists were the first to diagnose this crisis shows that knowledge is becoming deeply interdisciplinary and our understanding holistic. The 19th-century models of knowledge-building and siloed research, which the university is still built on, are a legacy and a hangover of how this knowledge originated and developed. So, if I were the praying type, I would pray to the planet to be patient with us: We are actually getting to know you and to relate to you.

Vinita Damodaran is a historian of modern India at the University of Sussex, where she is director of the Centre for World Environmental History. Her research focuses on sustainable development dialogues in the global South, particularly as it relates to questions of environmental change, identity, and resistance in Eastern India. Her work ranges from the social and political history of Bihar to the environmental history of South Asia, including using historical records to understand climate change in the Indian Ocean World. She is the author of several books and articles in established journals.

+ HISTORY, ENVIRONMENTAL JUSTICE

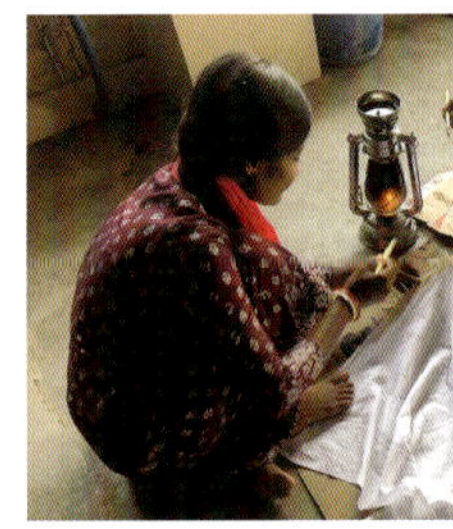

VINITA DAMODARAN

WHY

As a field within the discipline of history, environmental history has been described by Donald Worster, an early pioneer, as the interdisciplinary study of the relations of culture, technology, and nature over time. According to Richard Grove, it is the historically documented part of the story of the life and death of societies and species in terms of their relationships with the world around them. In this essay, I will discuss how a range of fears, anxieties, and disquiets about climate and environmental degradation in the early modern period, sparked by the destruction wrought by colonial empires, led to debates and initiatives for environmental preservation and conservation. Clearly, several of these concerns explored by environmental historians are relevant to our current dilemmas regarding the Anthropocene and global warming.

The European intellectual engagement with the environment and concerns about famines, diseases, and forest desiccation became particularly critical in shaping how modern Western colonial empires crafted their strategies for domination and expansion. For example, the British Empire, from the 18th century onwards, collected a veritable doomsday book of data on soils, weather, agriculture, geology, botany, zoology, and anthropology. Similar documentary pursuits were undertaken by other European empires, who also set about developing institutions for monitoring and conserving ecologies through a vast number of scientific networks that were spread across different colonies. In terms of climatic knowledge, the experiences of Europeans in tropical environments led to debates about extinction, endemism, and a nascent climatic environmentalism leading to draconian forest legislation. At the same time, the tropics were culturally perceived as leading to degeneracy and racial inferiority.

ENVIRONMENTAL

HISTORY

MATTERS

How can these pasts inform our present? How can we engage in a meaningful and productive dialogue about the relevance of early modern environmental histories, particularly forest histories of empire, to our current concerns about global warming and radical climate change? Here, I explore the significance of this history within the context of the emerging discourse on the Anthropocene, an era characterized by significant human impact on the planet. The Anthropocene has been described as "a dangerous and unpredictable era when fossil-fueled ways of life are undermining the planetary life-support systems upon which human societies depend."[1] Yet, these debates are not new. In his acclaimed monograph *Green Imperialism* (1995), Richard Grove traced the origins of modern Western environmental thinking to varied European encounters with Island ecologies and tropical environments between the 17th and 19th centuries. Modern environmentalism, Grove went on to argue, was not only a product of colonialism and European expansion but also emerged from the ways in which experiences in the environmental "margins" created a sense of crisis similar to what we experience today, forcing the Western world to reconsider its relationship with Nature through ecological sciences, forest conservation, and ethics of stewardship.[2]

Why does the history of colonial environmentalism matter to us today? The answer lies in the fact that the Indian forest-reserve system, developed under British rule, conserved one-fifth of India's landscape under forest reserves in response to the large-scale famines of the 19th century and fears of desiccation. This system became the basis for state land-use reservations on an enormous scale worldwide.[3] Categorized as reserved, protected, and village forest, the demarcation of the Indian forest system in 1878 was state-sponsored under a vast bureaucratic regime run by the grand Imperial Forestry Research Institute in Dehradun. It led to an exclusionary forest system in large parts of India, characterized by stringent forest reserves, which have caused chronic social conflict since their establishment. Though various conservation movements developed after 1870 in the temperate northern countries, within highly urbanized and prosperous societies, the preservation of wilderness–in the form of national parks or game reserves by the state, or the preservation of erstwhile common land–appealed mainly to reasons of aesthetics and species protection concerns. These reasons were sometimes applied to the tropics, especially in post-war Africa, but they were not as significant as the forest-reserve systems that developed in India.

Today, the planetary scale has become central to our understanding of contemporary climate change challenges. While climate change focuses on the problems of global warming, the critical relationships between the politics of carbon reduction and the environmental histories of forestry, hydrological cycles, ocean currents, and biodiversity loss await clearer articulation within climate change debates.[4]

Scientists are increasingly referred to as "historians of the future" due to their studies of planetary changes. However, environmental historians argue that locality, class, gender, and race are often overlooked in these discussions. This essay uses the lens of environmental history and ecological anthropology to understand long-term state responses to environmental change and community engagement with broader debates on resilience and sustainable development within the context of the Anthropocene. The climate crisis has led to a growing collaboration between conservationists and corporations, which has been shown to marginalize the local and Indigenous communities. Addressing this issue requires reintroducing historically nuanced, community-based histories into discussions of climate change and resource conflicts, thereby offering new political solutions to environmental crises. Examining the local impacts of the Anthropocene requires developing a more pointillist approach, necessitating targeted historical research. This approach calls for new emancipatory narratives that build on bottom-up approaches to our current crisis and on a sense of locality, community, and history. Social scientists and humanities scholars need to query the constructions of Anthropocene debates by scientists through a political examination of colonialism, capitalism, democracy, and wealth distribution. The shifting politics of climate change debates from "historical responsibility" to "we are all in it together" highlights the need for reframing these debates. It demands fresh thinking on the Anthropocene through a radical new political imagination to address resource wars, the militarization of boundaries and borders, and what one writer has called a discourse of a "good Anthropocene" emerging from a "bad Anthropocene."[5]

As leading historian Dipesh Chakrabarty has argued, the concept of the Anthropocene has become a key starting point for rethinking the humanities in the era of anthropogenic climate change. He raises key questions for historical research in the present and for the way we think historically–in terms of geological time and historical time–marking the end of the binary distinction between humanist and scientific knowledge that underpinned the Enlightenment rationalist project.[6] Chakrabarty suggests that the distinctive political project of the Enlightenment, the pursuit of human freedom (exemplified in the form of historical consciousness that enabled both liberalism and Marxism to function as political projects), is no longer tenable, and the debate on the Anthropocene underlines the need for a radically different kind of politics. How do we construct this new kind of emancipatory politics? Implicit in this politics is the relationship between capital and community, as well as a critique of capital through the recuperation of the history of the community and the local. One way forward is to write an environmental history of global capitalism, by understanding the impact of global capital on locality, community, and landscape within a historical time frame.

Resistance is at the center of this emancipatory project. The current debate on the Anthropocene is closely tied to debates on resilience and security. We need to view resistance and resilience not as antonyms, but as synonyms – resistance as a means by which people change social processes and build alternatives. However, there is little understanding of how communities have historically coped with natural disasters, such as floods, famines, and drought. How do people understand resilience? How is resilience mediated, and how does the power of community and place shape resilience? We need to understand everyday forms of resilience through human agency, collective action, and knowledge. The challenging of neo-liberal agendas by local communities resistant to maldevelopment allows us to understand resilience recuperated through resistance. We must reintegrate history and politics into resilience thinking and focus on agency and power in resilience science. Contemporary examples–ranging from resilience in the aftermath of the New Orleans disaster to fishing communities in the Philippines, Mexico, and Bolivia–have shown that communities less engaged with the market were more resilient to climate vulnerability.[7]

There is a need to assemble these hopeful narratives about the future, given the overwhelming belief that humanity is hurtling toward a mass extinction event. That is, we must challenge the pervasive "doom and gloom" mood by exploring human strengths for ingenuity and creativity. The intensification of disaster-related and catastrophic events such as flooding, droughts, and wildfires necessitates a considered and careful discussion of how states, communities, and cultures have handled extreme weather shocks and events in the past.

Eastern India: Forest Reserves, Mining, and Communities

Rapid industrialization and mining in Eastern India, particularly in regions like Odisha, Chhattisgarh, and Jharkhand, have profoundly impacted Indigenous communities. My research has shown that these communities have long suffered displacement under exclusionary colonial systems of forest reservation, established since the 19th century. The tribal rebellions of the 19th century forced the British to relax some of the forest rules and grant access to their resources under the Chota Nagpur Tenancy Act of 1908.[8] However, these communities face dual threats today, as these forest reserves are being diluted to accommodate mining interests, further eroding their natural resources. Since India's economic liberalization in the 1990s, mining activities have led to extensive pollution and resource conflicts. Indigenous communities, commonly referred to as Adivasis, have faced displacement and environmental degradation as their land is exploited for industrial projects. Despite their traditional conservationist practices, they are caught in the crossfire between development and environmental sustainability.

The concept of sustainable development, introduced in the 1987 Brundtland Report, has had little impact when applied in its current form. The environmental costs of development are disproportionately borne by the Global South, often in the form of pollution, displacement, and resource extraction. However, Indigenous communities in Eastern India offer an alternative model of sustainability rooted in intergenerational environmental stewardship. These communities have historically viewed their relationship with nature as one of coexistence, offering valuable lessons for global environmental debates.

The Saranda Forest in Jharkhand serves as a critical example of how Indigenous traditions clash with development interests and ontologies "that are not and never were our own."[9] This forest of *sal* (*Shorea robusta*) trees, the largest reserve in Asia, is extremely important in biodiversity and cultural terms both for local communities and the Indian imaginary. Local communities had long preserved their forest, and the ritual traditions of the sacred grove had clearly defined prohibitions against cutting down trees. In response to fears of land desiccation, colonial legislation was enacted, resulting in top-down models of conservation that excluded these communities. However, following the aforementioned tribal rebellions, the authorities relaxed forest regulations while protecting what they regarded as the "tribal natural heritage." These reserved forest laws went a long way in safeguarding significant forest cover in South Asia despite causing social conflict. However, the situation was to become much worse in the 2000s. In 2005, the Saranda forest, which is part of the core elephant reserve, was considered for World Heritage status by the Ministry of Environment and Forest. Later, it became a Maoist stronghold in the prolonged conflict between disaffected intellectuals, local communities, and the state over rights to the region's mineral resources and the growing poverty and inequality. Following the purge of Maoists from the region, the Saranda development plan sought to expedite the carving up of the reserve into mining leases. The mining industry in Eastern India, particularly in the aluminum, steel, and coal sectors, has generated significant debate over its environmental and social costs. While it has contributed to economic growth and job creation, it has also caused widespread pollution, displacement, and cultural disintegration. Indigenous communities face a "resource curse," wherein the wealth generated by natural resources benefits outside interests at their expense.

The story of Saranda Forest replicates the story of many protected and reserved forest areas in India today that are under threat from development. The displacement of Adivasi communities due to mining projects has transformed their way of life, shifting them from self-sustaining agricultural systems to industrial labor in resettlement areas. This transition has deepened poverty and eroded their cultural and ecological ties to the land. The Indian constitution's Fifth Schedule, intended to protect tribal lands, has often failed in practice due to loopholes that allow land acquisition in the name of public interest.

Resistance movements have emerged locally and internationally, with Indigenous groups and NGOs challenging the exploitation of their resources. Over 40% of environmental movements worldwide are led by Indigenous communities, giving them a significant role as environmental defenders.[10] These struggles reflect broader conflicts over access to land and environmental justice. Drawing on Indigenous sustainability practices, these movements in Eastern India aim to reclaim local agency and push back against the monopolization of natural resources. They build on conceptualizations and cultural origin myths, in which important deities are believed to be present in the distinctive mountain and deltaic landscapes, especially in the region's sacred woodlands (saunas or deswalis). Increasingly, Adivasi organizations assert their "minority" identities by promoting landscapes and historical sites of significance in resistance history as places of secular historical importance that warrant state protection. Because the resource frontier impinges on Adivasis, an understanding of economic choices induced by globalizing resources is enmeshed in the issue of the Adivasi's future. As Lewis and Maslin note in the global context of massive environmental and climate change:

> The power that humans wield is like any other force of nature and can be used withdrawn or modified. ...The recognition that human actions are driving far reaching changes to the life-supporting infrastructure of the earth...has philosophical, social, political and economic implications.[11]

Invoking recognition through their environmental resistance movements is vital to many of these communities and has important political and philosophical connotations. In recent children's paintings in the sinking Sundarbans delta, local children—some of whom are migrant Adivasis—depict their visceral understandings of their landscape and small-scale solutions to preserve their disappearing worlds. Their paintings represent a striking iconography of climate change and have the power to challenge, move, and demand a response from us.[12] Philosopher Charles Taylor notes that the politics of recognition "looms large in contemporary politics. From nationalist movements to demands on behalf of minority or subaltern groups in feminism and multiculturalism, the invocation of recognition is a mainstay of politics discourse."[13] We need to embrace this politics of recognition. Recognizing the religious and cultural worlds of the Adivasi—and acknowledging that our economic choices are enmeshed in the future of Adivasi communities—is critical, as is the recognition that our actions are driving far-reaching changes on Earth. In this context, understanding the histories of particular communities and their experience of ecological pressures and change over time is essential. This can be part of a broader social agenda aimed at local empowerment and environmental awareness from which emancipatory imaginings can emerge.

This essay has called for an interdisciplinary approach, combining environmental history and social anthropology. This approach examines long-term changes in landscapes and livelihoods while taking into account local cultural traditions and community responses. In Eastern India, key historical periods—including colonial interventions, post-independence development plans, and the economic reforms of the 1990s—have shaped the current environmental and social landscape. By studying archival records and oral histories, researchers can better understand how local communities have adapted to environmental changes over time. This knowledge is essential for developing alternative models of sustainable living that challenge the consumption-driven models of the West. If countries like India and China continue to emulate Western patterns of growth, the planet will face increasingly severe ecological consequences.

This research emphasizes the need to rethink sustainability through the lens of Indigenous traditions and local knowledge. By examining the environmental history of regions like Eastern India, we can uncover alternative ways of living that prioritize harmony with nature. While Indigenous practices should not be romanticized, they offer valuable insights into sustainable development and resilience. This essay advocates for a new political vision that draws on these traditions to address the challenges of the Anthropocene and build a more equitable future. Private capital and international trade interests have hitherto discounted environmental well-being and are only now being forced to consider these factors. The kind of climatic anxiety encapsulated in the recent intentions of international treaties on global warming has been a concern for centuries. History shows that only serious state intervention to control capital or corporate interests can hope to deal with the deadly risks of climate change, forest loss, and species extinction. Most importantly, only local communities, in effective collaboration with national governments and global environmental networks and institutions, can be guarantors of our fragile Earth's well-being for future generations.

1 Johan Rockström, et al., "A Safe Operating Space for Humanity," *Nature* 461, no. 7263 (2009): 475–79; Will Steffen, et al., "Trajectories of the Earth System in the Anthropocene," *PNAS* 115, no. 33 (2018), https://www.pnas.org/doi/epdf/10.1073/pnas.1810141115.

2 Richard H. Grove, *Green Imperialism: Colonial Expansion, Tropical Island Edens and the Origins of Environmentalism, 1600–1860* (Cambridge University Press, 1995).

3 Grove, *Green Imperialism*; Gregory Barton, *Empire Forestry and the Origins of Environmentalism* (Cambridge University Press, 2002).

4 Rob Allan, et.al, "Toward Integrated Historical Climate Research: the Example of Atmospheric Circulation Reconstructions over the Earth," *WIREs Climate Change* 7, no. 2 (2016): 164–74.

5 Simon Dalby, "Framing the Anthropocene: The Good, the Bad and the Ugly," *The Anthropocene Review* 3, no. 1 (2015).

6 Dipesh Chakrabarty, "The Climate of History: Four Theses," *Critical Inquiry* 35, no. 2 (Winter 2009): 197–222.

7 Katrina Brown, *Resilience, Development and Global Change* (Routledge, 2015).

8 Vinita Damodaran, "Indigenous Agency: Customary Rights and Tribal Protection in Eastern India, 1830–1930," *History Workshop Journal* 76, no. 1 (Autumn 2013): 85–110, https://doi.org/10.1093/hwj/dbs037.

9 Patrick Vinton Kirch & Marshall Sahlins, *Anahulu: The Anthropology of History in the Kingdom of Hawaii*, Vol. 1 (University of Chicago Press, 1994), 25.

10 Personal communication with Joan Martinez Alier, author of the Global Atlas of Environmental Justice, https://ejatlas.org/.

11 Simon L. Lewis & Mark A. Maslin, "Defining the Anthropocene," *Nature* 519, no. 7542 (2015): 171–80.

12 Anindita Saha & Vinita Damodaran, "The Mangrove School Project: Untold stories of children from the margins," in Francisco-J. Hernández Adrián & Angelos Theocharis (eds), *River Delta Futures: Endangered Communities in Audiovisual Media* (Bloomsbury, 2024).

13 Charles Taylor, "The Politics of Recognition," in Amy Gutmann (ed.), *Multiculturalism: Examining Politics of Recognition* (Princeton University Press, 1992).

Title Spread: Batik Art Workshop, Rajat Jubilee village, Chhoto Mollakhali Island, West Bengal. The Mangrove School Project.

JIP VAN BESOUW + LINO CAMPRUBÍ

MAKING THE SEASCAPE

Jip van Besouw is a historian and philosopher of science who works mainly on the physical sciences of the early modern period. He is currently a postdoctoral fellow at the University of Sevilla, in the project DEEPMED. Jip is interested in how environments were talked about and intervened in, historically. His most recent research focuses on conceptualizations and representations of rivers and seas in the 17th and 18th centuries.

Lino Camprubí is professor of History and Philosophy of Science at Universidad de Sevilla and PI of the ERC-CoG DEEPMED. He is the author of *Engineers and the Making of the Francoist Regime* (MIT, 2014), co-author of *Science and Apocalypse in Bertrand Russell* (Lexington Books, 2022), and author of "A Brave Deep World: Technologies in the Territorialization of Seas and Oceans" in the *Cambridge History of Technology* (Cambridge UP, 2025).

+ PHILOSOPHY OF SCIENCE, HISTORY

Amid rampantly rising temperatures and CO2 levels of our oceans, acidification of marine environments, and expanding offshore energy infrastructure, it is worth remembering that oceans were long considered unchangeable – spaces humans could not know or use. Historically, seas were generally perceived as the opposite of lands. This opposition was set out clearly by early modern legal scholars such as Hugo Grotius (1583–1645), who described the seas as naturally free. The *mare liberum* was outside of the realm of private property and politics, a space in which people and things had no impact. In contrast to lands, seas could not be built on, worked, plundered, or given shape. In short, Grotius held, seas could not be possessed or claimed and, therefore, ships of any nation should be allowed to move over them freely. Only *terra* could become territory.[1]

In this conception, oceans were not an "environment" with which humans could properly interact. Immutability meant immunity to impact. How did this perception give way to one in which oceans became spaces for humans to transform into "built environments"? Here, we set out how cultural and technological changes enabled new conceptions of seas as continuations of lands and, later, new metaphorical, material, and political uses of the seas.

It is important to note that Grotius's views were long dominant but never universally shared. First, there was resistance to the idea that immutability took oceans outside of politics: the early modern period witnessed a long debate about whether monarchs could impose *dominion* over the seas, for instance through naval warfare.[2] Besides, even those who followed Grotius in defending the doctrine of *mare liberum*, such as the 20th-century authoritarian theorist Carl Schmitt (1888–1985), implicitly granted political meaning to the ocean.[3] Like Grotius, Schmitt wrote to legitimize the dominant political powers of his time and, in fact, defended specific maritime political rights, ranging from protecting fishing grounds and merchant fleets to sinking and plundering enemy vessels. Second, alternative voices described oceans as less-than-immutable spaces. Imaginations of underwater worlds had always existed, fueled by tales and tangible experiences of sailors and fishers. From sea gods and mermaids to shipwrecks and sunken continents, not everyone imagined the deep sea as empty. Nevertheless, the ocean remained largely inaccessible to humans in these alternative imaginations for the simple reason that–apart from the first few meters under the surface frequented by pearl and sponge divers–living humans had no business *within* the deep seas. When did this start to change?

Dikes, Coastal Defenses, and Rising Seabeds

It was through early modern investigations of rivers and shallow seas, directly related to constructing and controlling coastal landscapes, that new perceptions of deep seas arose. A first relevant case of conceptualizing how human intervention was changing the seas comes from the Dutch Republic. Dominated by the delta plain of the Rhine, Meuse, and Scheldt, the inhabitants of the Low Countries had built dikes at least since the 11th century. With these dikes, central to efforts to create and protect arable lands, the Dutch actively made their fluvial landscape.[4] In the 1720s and 1730s, one of the leading cartographers and meteorologists of the Republic, Nicolaas Cruquius (1678–1754), asserted that this reshaping of Dutch rivers impacted the seas as well. While doing so, Cruquius gave one of the first quantitative estimates in history of how fast sea levels were rising.[5]

This was not just empty speculation. Acting as a special advisor to governments, Cruquius was asked to investigate and report on water management at specific locations. His reports addressed issues ranging from coastal erosion and river sedimentation to the building of new canals. Dealing with the case of sedimentation at the mouth of the Meuse near Rotterdam, Cruquius came to the conclusion that sediment transport by the river was so significant that it would make the seabed rise locally, around the river mouth. Based on general observations on the Rhine basin, Cruquius claimed that the rate of sea rise could be calculated through an assessment of land erosion by rainfall, the amount of sediment suspended in the river, and the locations over which these sediments settled. Extrapolating from his observations of the Rhine basin, Cruquius speculated that sediments dumped into the sea by rivers worldwide could raise the global seabed by roughly one foot every two hundred years. As the water had nowhere to go but up, the sea level would rise accordingly, Cruquius suggested.[6]

This early assessment relied on a staggering overestimation of how much sediment is actually suspended in river water: present-day climate science tells us river sediments are a negligible factor in sea level rise. Notwithstanding, what is worth underlining is that Cruquius made the design of local infrastructures a fundamental part of his discourse on sea level rise. His assessment was that the changing situation of the Meuse would lead to flooding and problems with navigating the river in the short term. To prevent those problems, he called for the centralized management of rivers, dikes, and coastal defenses in the Netherlands. This plan to restructure Dutch water politics made a point specifically of standardization of the architecture of river dikes, with Cruquius pointing out that the failure of individual dikes would alter the distribution of waters, and thus sedimentation, along a large stretch of the river, including its mouth.[7] Cruquius thus saw dikes as a means of controlling the rise of seabeds, at least at a local level. Consequently, he turned the bottom of the sea into an environment that could be measured, controlled, and impacted.

The Architecture of the Earth and Sea

Two contemporaries of Cruquius investigated the seabed for entirely different reasons. Both Luigi Ferdinando Marsigli (or Marsili) (1658–1730) and Philippe Buache (1700–1773) were primarily interested in the underwater world to make claims about how the seas fitted into the making of our planet more generally. Marsigli, an aristocrat born into a leading Bolognese family, was a diplomat and former colonel of the Habsburg army, from which he was dishonorably discharged in 1704 after failing to hold a fortress on the banks of the Rhine against the French. In the years immediately afterward, Marsigli doubled down on the naturalist interests he had cultivated throughout his career.[8] Settling down in the domain of his former enemy, Marsigli made detailed investigations into the hydrology, geography, and marine life of the Languedoc coast and the Gulf of Lyons. Almost two decades later, Marsigli published his findings in his *Histoire Physique de la Mer* (1725). In its first two chapters, he presented his measurements of the waters' depths, density, and salinity. These investigations led Marsigli to claim that land-based mountain chains continue under the seas. He presented evidence for this claim in the form of profile maps showing continuous slopes from the lands into the seas, as well as by pointing out that the seabed consisted of materials similar to those of the lands bordering them. This led him to conclude "that the Basin of the Sea was formed during the Creation of the same stones which we see, in the layers of the Earth."[9] That is, Marsigli claimed that seabeds were an integral part of the planet, designed some five thousand years ago, just like the rest of the Earth.[10]

What was Marsigli's interest in bringing marine environments into the "Anatomy of the Earth"? In a diatribe against people who pursued knowledge only for practical ends such as navigation, Marsigli claimed that he had been led by his love of science. He wanted to understand how God had created nature, and particularly to show the order, works, and laws of nature thus established.[11] Illustrations of such order were that the river Rhone continued to flow over the seabed; that opposing coasts were alike in shape; and a suggestion, brought almost as a matter of fact, that the largest depths beneath the surface of the seas were similar to the highest mountain peaks above it.[12] Clearly, Marsigli saw his book as the study of the structure God gave to the Earth, what we could call the supposed divine architecture of the Earth.

This interest in the order of creation was, in more than just a metaphorical way, related to the order of politics.[13] Besides being a formerly high-ranking military officer, Marsigli was a member of some of the most elite societies of science in Europe, including the royally sponsored Academy of Sciences of Paris, to which he dedicated his book. When Marsigli asked for a "prince," a "protector of science," to fund more systematic studies into the seas, this was by no means an empty call.[14]

Links between geography and politics are even clearer in the work of Buache, who became the Royal Cartographer of France in 1730 and was paid by the same Academy of Sciences of Paris to map all parts of the globe. Buache's theory on the continuity of lands and seas went further than Marsigli's: Buache claimed that the architecture of mountain ranges above and underneath the sea level was what kept the Earth's surface in place. He leveraged this view to intervene in debates on how to delimit administrative regions in a supposedly natural way, to speculate on whether a North-West passage to Asia existed, and to predict the layout of yet unmapped regions of the world.[15] Thus, in contrast to Grotius, who saw lands and seas as opposites, Marsigli and Buache presented them as being geologically similar and continuous, and used that conception of the seas to inform global politics.

Building Underwater

Early measurements of depth, as well as interest in underwater geological features and biological creatures, turned into more systematic engagements with the deep seas in the 19th century. Submarine telegraphs were among the first fixed underwater structures. These cables were technologies of imperial control. First tried out in India and pioneered between Britain and France in 1850, submarine telegraphy developed at an incredible speed.[16] By 1854, the French, supported by the authorities of the Kingdom of Sardinia, had laid a cable between La Spezia on mainland Italy and Corsica, a distance of over 100 km and at a depth much greater than that of the English Channel. In 1855, a new section of this line connected Algeria, a French colony, reaching 500 fathoms, about 1 km deep.[17] The Mediterranean Sea, like other seas and oceans around the world, was soon traversed by signal-carrying lines.

These early attempts at submarine telegraphy were fraught with technical problems. The first cable to cross the Atlantic Ocean, between Ireland and Newfoundland, broke down less than one month after the first transatlantic messages had been sent in August 1858. Broken cables had to be brought to the surface to be repaired or entirely replaced by new cables. The brittleness of this submarine infrastructure directly inspired three sustained lines of new scientific investigation. First, hydrographers developed a range of measuring instruments that took bathymetry to the next level and allowed better decisions on where to lay cables.[18] Second, to make cables send strong enough signals over long distances and under the massive pressure of the ocean, investigations of the materials involved were scaled up. Research focused on electromagnetic theory and the resistance of gutta-percha, the material used to insulate the cables, to erosion.[19] Finally, engineers reported deep marine life attached to retrieved cables, and marine biologists certified the death of the azoic hypothesis, the idea that deep seas did not support life.[20] The deep seas were geologically and biologically as rich as land-based landscapes. The *seascape* thus emerged as a new frontier of knowledge and empire.

Other permanent and mobile technologies soon joined telegraphs in the making of oceanic built environments. Submarines and underwater mines were central to naval battles in World War I, and detecting them became a strategic imperative. Military powers invested in both passive and active sonar, which could listen to sounds emitted by enemy vessels and echo-locate solid objects, including the bottom of the sea.[21] As the propagation of sound waves depends on the density of the medium—often measured through proxies such as salinity and temperature—accurate detection had to consider the properties of the different layers of seawater. Navies around the world moved toward knowledge and surveillance of the seas and seabed.[22]

Acoustic technologies also impacted conceptualizations of the geology *under* the seabed. Turning around the earlier logic of using properties of the medium to understand the propagation of sound waves, toward World War II, sound waves became central to studying sediments, thus picking out promising places for oil drilling. Offshore oil investors changed the architecture of the seas entirely, as platforms and artificial islands, supported by underwater columns of concrete, were constructed to aid oil extraction.[23] Together with the invention of the scuba bottle to transport oxygen under high pressure, the time seemed ripe for a full-scale invasion of the ocean by humans. By the 1960s, advocates of bioengineering came up with various outlandish projects for making the deep sea fit for human habitation, from sea labs in which researchers could spend long spans of time to

1 Hugo Grotius, *The Free Sea, Natural Law and Enlightenment Classics*, trans. Richard Hakluyt, ed. David Armitage (1609; repr. Liberty Fund, 2004).

2 Nieves San Emeterio Martín, "El debate sobre el dominio de los mares en el imperio español durante los siglos XVI y XVII," *Iberian Journal of the History of Economic Thought* 7, no. 2 (2020): 133–42.

3 Carl Schmitt, *Land and Sea: A World-Historical Meditation*, trans. Samuel Garrett Zeitlin, ed. Russell A. Berman & Samuel Garrett Zeitlin (1942; repr. Telos Press, 2015).

4 Petra J. E. M. van Dam, "Sinking Peat Bogs: Environmental Change in Holland, 1350–1550," *Environmental History* 6, no. 1 (2001): 32–45, https://doi.org/10.2307/3985230.

5 On sea level rise more generally, see Wilko Graf von Hardenberg, *Sea Level: A History* (Chicago University Press, 2024).

6 Nicolaas Cruquius, "Korte verhandeling, van de opkomst; verandering; tegenwoordige en toekomende staat, van 't Vaderlandt ten opzichte, van de wateren" (May 30, 1731), Bijzondere Collecties, KW 129 C 8 [4], Koninklijke Bibliotheek, The Hague, 9. This piece is discussed in Mathijs Boom & Jip van Besouw, "Rising Seas, Sinking Lands: Reckoning with Local and Global Sea Level in the Early Modern Netherlands" (ms).

7 Nicolaas Cruquius, "Eenige Aenmerkingen over de Waterstaet Van dese Landen" (ca. 1726), Archief van Hoornebeek, no. 309: [f. 1r], National Archives of the Netherlands, The Hague.

8 On Marsigli's career, see Anita McConnell, "L. F. Marsigli's Visit to London in 1721, and His Report on the Royal Society," *Notes and Records of the Royal Society of London*, 47 (1993): 179–204, https://doi.org/10.1098/rsnr.1993.0026.

9 Louis Ferdinand Marsigli, *Histoire physique de la mer* (Amsterdam, 1725), p. 14: « on peut conclure ... que le Bassin de la Mer fut formé dans la Création de la même pierre que nous voyons, dans les couches de la Terre.»

10 Ibid., 39.

11 Ibid. See the two prefaces to the book.

12 Ibid., 11. As Marsigli's depth measurements did not go beyond a few hundred meters, his suggestion that depths of seas equaled heights of mountains was entirely without evidence.

13 The importance of nature as a metaphor for social and political order should not be underestimated, especially in the early modern period. For an introduction, see Lorraine Daston & Fernando Vidal (eds), *The Moral Authority of Nature* (Chicago University Press, 2003).

14 Marsigli, *Histoire physique de la mer*, 47.

15 Michael Heffernan, "Geography and the Paris Academy of Sciences: Politics and Patronage in Early 18th-Century France" *Transactions of the Institute of British Geographers, 39* (2014): 62–75, https://doi.org/10.1111/tran.12008.

16 Daniel R. Headrick & Pascal Griset, "Submarine Telegraph Cables: Business and Politics, 1838–1939," *Business History Review* 75, no. 3 (Autumn 2001): 543–78.

17 Roberto Mantovani, "The Otranto-Valona Cable and the Origins of Submarine Telegraphy in Italy," *Advances in Historical Studies* 6, no. 1 (2017): 18–39.

oxygen-feeding "cyborgs"–humans who would be engineered so that they could live permanently on the seabed.[24]

Real and imagined infrastructures, together with legal instruments, opened the door to the territorialization of seas and oceans, thus putting an end to *terra*'s monopoly over territory.[25] Initial political moves to control the exponentially growing economic and military potential of the seas harked back to the view of seas as continuations of lands: in 1945, the United States claimed sovereignty over the seas in front of its coasts, as far as the continental shelves reached. Building on a loose geological conceptualization, other countries followed suit and claimed sovereignty over "their" continental shelves, as if they were "submerged lands."[26] Legal negotiations followed. The prospects of mineable manganese nodules on the seabed's surface, depleting fishing stocks, and submerged nuclear deterrence led to various United Nations Conferences on the Law of the Sea. A treaty signed in 1982, and still active, allows states to claim an Exclusive Economic Zone stretching to 200 nautical miles off the coast. In these zones, states have exclusive rights to exploit marine resources, or to sell these rights to large corporations.[27]

Conclusion

Seas are not lands. They are different on an ontic, or material level: seas are liquid, and they do not react physically like the solid earth or the gaseous atmosphere do. Nevertheless, through science and technology, humans have greatly transformed this underwater world to support our activities on land. The internet depends as much on a global network of underwater cables as on satellites shot into space. Pipelines flushing oil and gas continue to fuel the global economy. The global economy, in turn, continues to demand more from the underwater world. More than anything else, the importance of power–both political power and energy infrastructure–has turned the oceans into human, built environments. Although seas and lands remain distinct in many important ways, as a political space, seas have become increasingly similar to *terra*. Unfortunately, the increasing extraction of their riches by states and large corporations goes hand in hand with their destruction. The oil and gas flowing through pipes at the bottom of the seas drive the warming and acidification of the global ocean, while the pollution that can be expected from deep sea mining will surely accelerate environmental breakdown. Although not for the correct reasons, Cruquius was right: humans can and do alter the sea level and other basic features of the seascape by their action as well as by their inaction.

18 Helen Rozwadowski, *Fathoming the Ocean: The Discovery and Exploration of the Deep Sea* (Harvard University Press, 2005), 69–95.

19 Bruce J. Hunt, *Imperial Science: Cable Telegraphy and Electrical Physics in the Victorian British Empire* (Cambridge University Press, 2021).

20 Lino Camprubí, "'No Longer an American Lake': Depth and Geopolitics in the Mediterranean," *Diplomatic History*, 44, no. 3 (2020): 428–46.

21 Robert K. Massie, *Castles of Steel: Britain, Germany, and the Winning of the Great War at Sea* (Ballantine Books, 2003); Willem Hackmann, *Seek and Strike: Sonar, Anti-Submarine Warfare and the Royal Navy, 1914–54* (Her Majesty's Stationery Office, 1984), 45–71.

22 Sabine Höhler, "Depth Records and Ocean Volumes: Ocean Profiling by Sounding Technology, 1850–1930," *History and Technology* 18, no. 2 (2002): 119–54; Naomi Oreskes, *Science on a Mission: How Military Funding Shaped What We Do and Don't Know About the Ocean* (Chicago University Press, 2021); Lino Camprubí & Alix Hui, "Testing the Underwater Ear: Hearing, Standardizing, and Classifying Marine Sounds from World War I to the Cold War," in Viktoria Tkaczyk, Mara Mills & Alexandra Hui (eds), *Testing Hearing: The Making of Modern Aurality* (Oxford University Press, 2020), 301–25.

23 Jon Phillips, "Order and the Offshore: The Territories of Deep-Water Oil Production," in Kimberley Peters, Philip Steinberg & Elaine Stratford (eds), *Territory Beyond Terra* (Rowman and Littlefield, 2018), 51–68.

24 Helen M. Rozwadowski, "Engineering, Imagination, and Industry: Scripps Island and Dreams for Ocean Science in the 1960s," in Helen M. Rozwadowski & David Van Keuren (eds), *The Machine in Neptune's Garden: Historical Perspectives on Technology* (Canton, 2004), 325–52.

25 Lino Camprubí, "A Brave Deep World: Technologies in the Territorialization of Seas and Oceans," *Cambridge History of Technology, vol. 3.* (in press).

26 Donald C. Watt, "First Steps in the Enclosure of the Oceans: The Origins of Truman's Proclamation on the Resources of the Continental Shelf, 28 September 1945," *Marine Policy* 3, no. 3 (July 1979): 211–24.

27 Helen M. Rozwadowski, "Wild Blue: The Post-World War Two Ocean Frontier and Its Legacy for Law of the Sea," *Environment and History* 29, no. 3 (2023): 345–76.

Acknowledgment
Research for this paper was funded by the research grant ERC-CoG-DEEPMED-101002330.

ANDREW WITT

SNOW GLOBES: CYCLIC DRAWING AND CRYOSPHERIC PANORAMAS

Andrew Witt is an associate professor in practice at the Harvard Graduate School of Design, teaching and researching on the relationship of geometry, data, AI, and machines to architecture, design, and culture. He is also co-founder, with Tobias Nolte, of Certain Measures, a Boston/Berlin-based studio that combines design and data for systemic and scalable approaches to spatial problems.

+ ARCHITECTURE, VISUAL ARTS

In his sensational memoir, *The Worst Journey in the World*, English explorer Apsley Cherry-Garrard recounts the fateful and harrowing 1910 *Terra Nova* Antarctic expedition. The expedition, best known for the dramatic deaths of Captain Robert Falcon Scott and his party, also had broad geographic and scientific ambitions. Cherry-Garrard's volumes, which record scientific excursions to collect diverse geological and biological samples, are animated by the haunting watercolor images and meticulous sketches of landscapes by expedition artist Edward Wilson. Among the most remarkable is Wilson's panoramic drawing off the coast of Ross Island, a 360-degree view of the end of the Earth. This drawing, a testament to Wilson's dedication and precision, opened the eyes of the British public to this remote and otherworldly environment for the first time.

Wilson's panoramas were a transformative convergence of media and natural environments, offering a uniquely immersive experience. These panoramas, along with later technologies such as stereoscopy and cinema, signaled a 19th-century

transformation in the concept of the environment. Drawing a panorama effectively distilled and encoded a natural environment into an artificial media simulation. Wilson's panoramas thus represent the virtualization of the natural world onto a technical sensoria that is a spatial environment, offering viewers an immersive and transportive experience that reshapes their perception of the icy world.

While unusual to the modern eye, Wilson's panoramas were not so to his colleagues on the *Terra Nova*. They would have recognized them as fitting squarely in the British tradition of military reconnaissance panoramas. This tradition, a forgotten but essential tool to surveil the front lines of conflict since the Napoleonic wars, involved the execution of panoramic drawings by specially trained (and undeniably intrepid) draftsmen. These draftsmen would furiously sketch in a perilous frontline zone, rushing their drawings back to plotting commanders stationed at a safe distance. Reconnaissance panoramas were a tactical tool, replicating a hostile environment through a specific lens. Historian Paul Gough notes that "gentlemen cadets and sailors were trained to analyse and record landscape and coastline as a means of neutralizing and controlling enemy space," scrupulously capturing any hillock or glade that could mean the difference between glorious victory or ignominious death.[1] With no time for frivolities, this peculiar drawing aimed to be unabashedly objective, a detached and faithful recording of the site. Indeed, some artists dismissed the panoramacists as practicing mere "tame delineation" rather than genuine art.[2] Yet the techniques suited the rugged extremes of polar exploration ideally. Since military officers often staffed polar expeditions, it should be of little surprise that they imported their techniques of seeing to a new campaign of conquest, this time at the edge of the world.

Striking panoramas like Wilson's were recurring representations of the desolate frozen environments of the remote north and south during the high age of polar exploration. Produced with varying aims, from scientific documentation to public education, these polar panoramas strove to capture the sublime feeling of vast and magisterial icescape through the eyes of an observer surrounded by nothing but snow and sky. Occasionally, they were scaled up to breathtaking architectural effects, leaving the fascinated public in awe and wonder.

About a century before the *Terra Nova* expedition, an eerily similar and equally dramatic scene—an ice-entombed mountainous landscape punctuated with the craggy outcroppings of crystalized water—unfurled on just such a grand architectural panorama. In 1819, panoramic painter and entrepreneur Henry Aston Barker (1774–1856) presented "The North Coast of Spitzbergen," a stirring tableau of an 1818 British naval polar expedition, in his family's vast rotunda building in Leicester Square, London. Barker's father, Robert Barker (1739–1806), had invented the entire medium of panoramic painting "for the purpose of displaying views of nature at large," patenting his process in 1787.[3] This panorama used the medium to great effect, tracing a horizon punctuated by cathedrals of ice thrust skyward and the masts of hulking vessels as they surveyed the Hyperboreal territory.

The narrative text that accompanied the spectacular view of Spitzbergen reads like a passage torn from Cherry-Garrard: "From the violence of the waves, immense sheets of ice, many hundred thousand tons weight, were tossed about in all directions, or hurled against one another. Floes, of several acres in extent, were rent asunder, or crumbled to atoms."[4] The extraordinary drama "presented a scene of horror far beyond the power of language to fully describe."[5] The strange, indescribable scene enraptured the public. In a letter to his brother George dated April 21, 1819, the poet John Keats rhapsodized that "I have been very much pleased with the Panorama of the ships at the North Pole—with the icebergs, the mountains...it is impossible to describe the place."[6] In this polar panorama, a bodily intensity of reality met a sublime illusion that eluded words.

Contemporary visitors marveled at the near-perfect reality of the fastidiously detailed painting and its seamless continuity as it surrounded them. To achieve this remarkable illusion, Barker instructed the draftsman to synchronize the hand with a cyclic movement of the entire body. In his patent text describing the technique of panoramic drafting, Barker relays that "the painter or drawer must fix his station, and delineate correctly and connectedly every object which presents itself to his view as he turns round, concluding his drawing by a connection with where he began."[7] In effect, the draftsman enacted through his hand the sweeping motion of the eye, continuously surveying the land.

Beyond the virtuosic representational technique itself, the peculiar force of the Spitzbergen panorama owed much to the architecture in which it was mounted: a grand purpose-built rotunda comprising a cylinder of 90-foot radius and 57 feet high.[8] For Robert Mitchell, the architect of this unusual rotunda, the effect of the panorama was to create "so true a representation of the appearance of natural objects" that it achieved an almost hallucinatory verisimilitude. An unbroken illusion was possible thanks to the fact that "the whole circle of the horizon is seen without interruption," seamless and endless. This poetic image—the line of the horizon bent back on itself, ouroboros-style, in a continuous and infinite round—distilled the geometric essence of the panorama. In his building, constructed precisely to Robert Barker's specifications, Mitchell claimed that "the Panorama has arrived at a perfection to render the exhibition of it a source of pleasing and rational amusement to the Public."[9] The rotunda became a vast optical instrument, tuned to the scale of the human body, to create an experience that nearly surpassed reality itself.

The Leicester Square Rotunda was the prototype of a typology of architectural panoramas that would proliferate across Europe and America, reaching its apex around 1900 and declining with the advent of film projection spaces in the early 20th century. The golden age of the panorama coincided with the high point of polar and alpine exploration. So, the polar panorama constitutes an appropriate image genre to showcase the unusual powers of panoramic architecture. Yet beyond dramatic polar scenes, vast panoramas captured various other epic scenes, including grand military melees, pitched naval battles, and intricate urban vistas. Views of Edinburgh, London, and Bath were among the first panoramas to be shown by Barker.[10] These two thematic antipodes—the civilized metropolis and the icy wilderness—bring the viewer to the grandest manifestation of human designs on the one hand and the remote extremities untouched by human designs on the other through the familiar device of the panorama.

The unique representations that distilled these overwhelming spatial spectacles into drawn representations on the two-dimensional flat page were more intriguing still than the architectural panoramas themselves. Exhibitors achieved this through an ingenious anamorphic projection from the cylindrical elevation into a radial drawing called a "panorama key." The panorama reformatted the encompassing bodily experience of the architectural panorama into a memento that fit in one's hand. These keys had a strange, cryptic, almost magical aesthetic in which forms tumbled, landscape flowed, and architecture vertiginously pirouetted.

Panorama keys accompanied many exhibitions at the Leicester Square panorama and the numerous imitators and competitors that emerged throughout Europe in the 19th century. Panorama keys were maps to cylindrical elevations and indices to specific information embedded in the larger composition. They were interfaces between visitors and buildings, and they were also peculiar architectural representations. Though they share superficial similarities, panorama keys can be contrasted with their cousins – developed surface drawings.[11] They both flatten interior elevations using clever projections, but the panorama key is far more technically ambitious. While the developed surface drawing cuts seams between walls and folds each planar wall elevation onto a target plane by orthogonal projection, panorama keys flatten and distort elevations to preserve an integral shape without cuts, breaches, or divisions. The projection employed by the panorama key is not a simple unfolding but rather a differential remapping and stretching through continuous local deformations. By elastically reformatting a continuous perspective, the panorama key becomes globe-like, a world in itself. Mathematically, this projection is sometimes described as stereographic, a "compactification" of an infinite plane into a finite disc.

Of course, the cryosphere is by no means confined to the poles, nor are its panoramic representations confined to cylindrical projections and their derivatives. The cryosphere also touches the heights of alpine regions draped in glaciers or steppes blanketed in permafrost. Also, panoramic anamorphism was crucial for communicating a frozen land's ineffable mysteries. Inventive artists, scientists, and adventurers employed stereographic projections and similar techniques to flatten vast panoramic vistas into planar disc views directly, without the intermediary construction of the cylindrical panorama.

A particularly striking example of a stereographic panorama appears in the compendium of illustrations that accompanies *Voyages dans les Alpes*, a monumental essay on the natural history of polymathic Genevan scientist Horace Bénédict de Saussure (1740–1799). In this work, one sees early intimations of meteorology, hydrology, and glaciology as de Saussure transects the environment of the Alps, attending to the overlapping worlds of atmosphere, hydrosphere, and cryosphere. His representational innovations played a role in opening new ways of reading about the Alps and climate in general. Of de Saussure's drawings, one curious hand-drawn panorama, a circular sketch resembling a craggy rock sphere or rugged asteroid, is particularly arresting. The drawing evokes an inchoate vortex lent structure by the induced circularity of stereographic projection. Precisely foreshadowing Barker's patent text describing the technique of panoramic drawing, de Saussure's remarkable account of his technique connects the body to the landscape through representation:

> The draughtsman makes an exact copy of the objects as he sees them by turning the sheet as he himself turns, and anyone wishing to use his work to get an idea of what the object he has drawn looks like must imagine that they are at the centre of the drawing, and with their imagination enlarge on what they see from this centre and, turning the drawing, inspect all the parts. In this way, they will see how, one after another, all the objects are linked together and exactly how they appear to the observer who stands at the top of the mountain.[12]

The drawing induces a kind of circular orbit in the reader, a conceptual merry-go-round that faintly recalls the dizzying disorientation in the attenuated air of the alpine atmosphere.

To bring our panoramic survey full circle, on the frontispiece of the second volume of *The Worst Journey in the World*, E. A. Wilson limns a luminous and haunting halo effect known as a parhelion. Appearing unusually bright in polar climes, this strange ring-like optical artifact is caused by the lensing of light by ice crystals gently suspended in the atmosphere. Though parhelia typically appear as rings circumscribing the sun, in Antarctic regions, ice crystals are so thick in the air that even the faint crepuscular light of the moon can produce breathtaking halos. In his watercolor of a phantasmic "halo round the moon," Wilson conveys an ecstatic amplification of lunar light as if the horizon line of the waning sun was bent into an exact circle around it. In this strange landscape, we even see the moon with fresh eyes.

In his engaging study of the cultural history of the north, literary scholar Peter Davidson argued that in our collective imagination of the cryosphere, "the familiar images of hostile desert and place of austere truth are combined afresh."[13] With its implausible ice formations and uncanny parhelia, it is a "place of treasures and marvels...the site of prodigies," a world apart.[14] Cryospheric panoramas were techniques to capture and experience these prodigies elsewhere, portals for

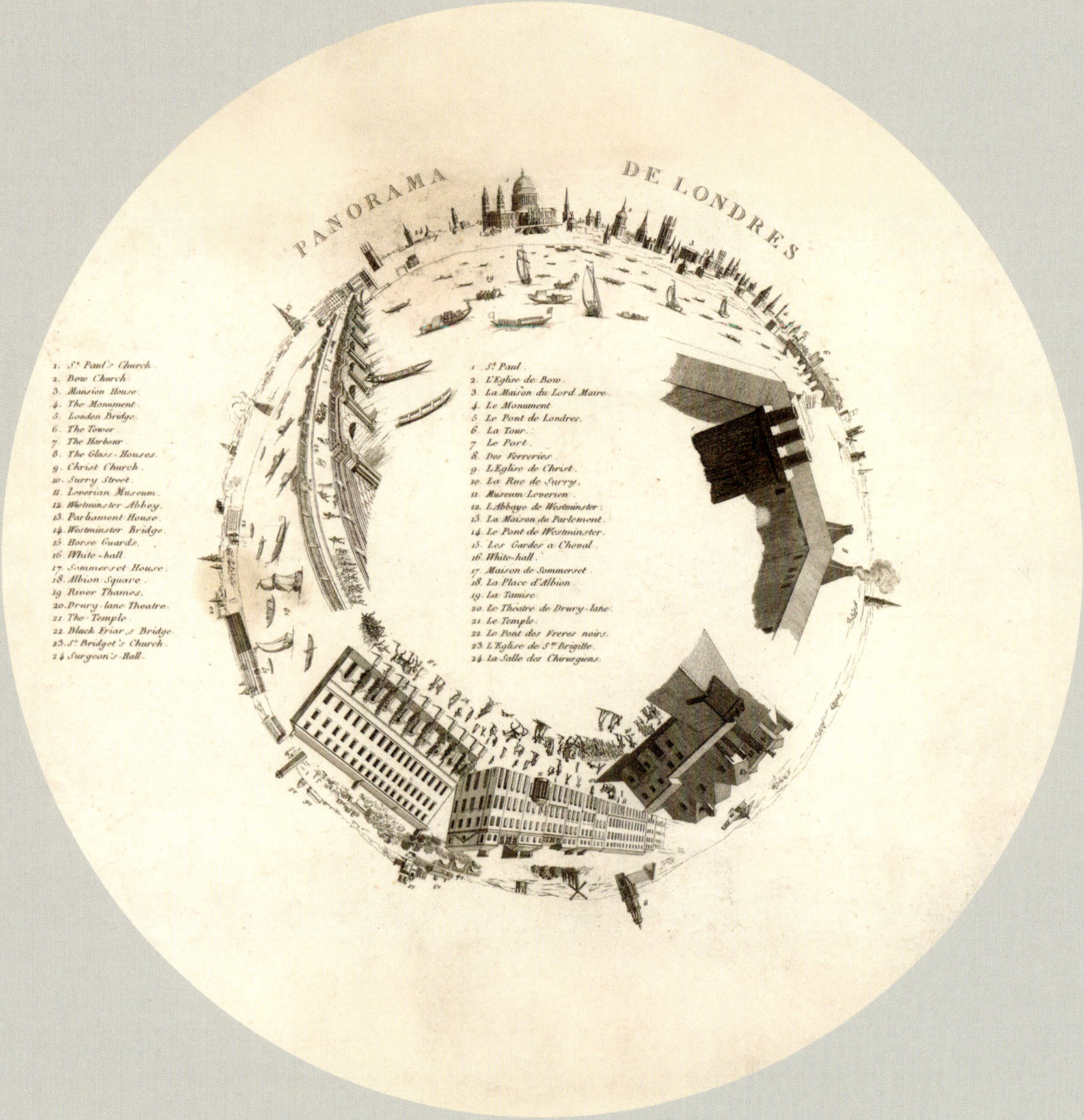

Above: The key of Barker's panorama of London, as produced for its exhibition in Paris. Image courtesy of the British Museum.

Above: Section of the Rotunda, Leicester Square, exhibiting a panorama of London from the River Thames by Robert Mitchell (1801).

translating observers into the alien regions of the far north and south. They offered a paradigm of climatic and environmental representation–often at the scale of architecture–at a time when those concepts were yet vague intimations.

A panoramic environment is a total sensorium, a closed representation. Yet beyond a sensorium, a panorama hints at the vast, veiled systems that undergird and generate those sensations – the invisible environments of laws, rules, relations, and ecological dynamics. A panorama is a dance between a media-representation and a natural-spatial environment – a complex fusion of these two with other nearly infinite environments: atmospheric, hydrologic, geologic, optical, cultural, economic, symbolic, and computational, to name a few.

Today, panoramic images of the cryosphere are an unintended act of preservation as much as representation. They memorialize an icy world in rude health to which we may not return, a set of treasures and prodigies that may soon slip into legend and myth. Through the lens of the panorama, they preserve not only that world but a lost way of drawing and representing worlds.

1 Paul Gough, "Calculating the Future: Panoramic Sketching, Reconnaissance Drawing and the Material Trace of War," in Nicholas J. Saunders & Paul Cornish (eds), *Contested Objects: Material Memories of the Great War* (Routledge, 2009), 237.

2 Ibid., 238.

3 *Repertory of Arts and Manufactures* (George and Thomas Wilkie and G. G. and J. Robinson, 1798), 165.

4 Henry Aston Barker, *Description of a View of the North Coast of Spitzbergen* (J. W. and C. Adlard, 1819), 6.

5 Ibid., 6.

6 Hyder Edward Rollins (ed.), *The Letters of John Keats, 1814–1821*, vol. 2 (Harvard University Press, 1958), 95.

7 *Repertory of Arts and Manufactures*, 165.

8 It contained two concentric panoramas, the second much smaller than the first, but it was on the larger one that the most monumental exhibitions appeared.

9 Robert Mitchell, *Plans, and Views in Perspective, with Descriptions, of Buildings Erected in England and Scotland: and also an Essay, to Elucidate the Grecian, Roman and Gothic Architecture, Accompanied with Designs* (Wilson & Co., 1801), 7.

10 Markman Ellis, "'Spectacles Within Doors': Panoramas of London in the 1790s," *Romanticism* 14, no. 2 (2008): 133–48.

11 This type of drawing unrolled an architectural surface onto a flat plane without stretching or distortion, thus representing the true shape of the surface in question. Ironically, such a drawing often splits the architectural surface through joints or seams, creating a discontinuous or fragmented projection. Historian Robin Evans popularized the term in the well-known essay "The Developed Surface: An Enquiry into the Brief Life of an Eighteenth Century Drawing Technique" in Evans, *Translations from Drawing to Building and Other Essays* (Architectural Association Publishers, 1997).

12 As quoted by William Slattery Lieberman in Bernard Comment, *The Painted Panorama* (Metropolitan Museum of Art, 2000), 82.

13 Peter Davidson, *The Idea of North* (Reaktion Books, 2005), 48.

14 Ibid., 50.

Opposite: Image by Edward Wilson from Apsley Cherry-Garrard, *The Worst Journey in the World*, Vol. 1 (1922).

DIANA GUO

MEGAFAUNAS OF THE PAST AND FUTURE

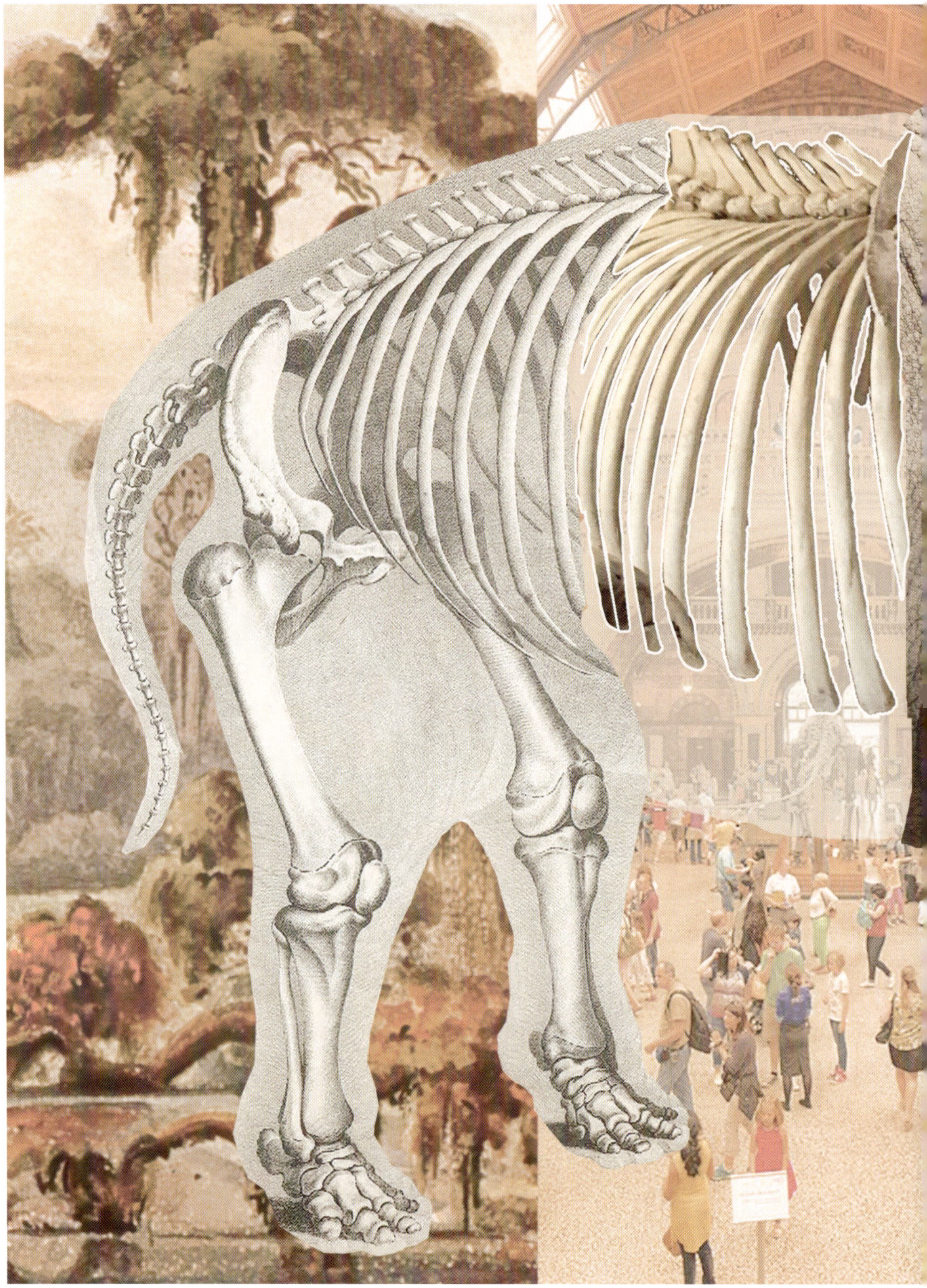

Diana Guo is a PhD student in urban planning at Columbia University GSAPP. Her research focuses on transpacific urbanisms, alternative forms of land stewardship, and the infrastructure of care in Asian immigrant communities. Working across design, critical theory, and research, her praxis is concerned with decolonial landscapes and cities. She holds a masters in landscape architecture and MDes from Harvard GSD with distinction, and a bachelor of art in urban studies from Vassar College.

+ NATURAL HISTORY, DESIGN

At the New York American Museum of Natural History, on display on the first floor of the Irma and Paul Milstein Hall of Ocean Life, a colossal *Balaenoptera musculus* (blue whale) is suspended in motion, diving in mid-air. First created in the 1960s, the 94-foot-long model was reconstructed based on photographs of a female blue whale corpse found in 1925 in South America. By the early 2000s, exhibit-makers and biologists had obtained much more information about blue whales through deep-ocean cameras and live feeds, so the big blue whale underwent surgery again: her large round eyes were made smaller, and her tail was sharpened. She even received a belly button.

According to the official website, her belly button can be found approximately "four-fifths of the way down the whale, a reminder of just one of the traits humans share with this majestic mammal."[1] Although appearing lifelike in corporeal form, the suspended blue whale is a technological invention—an amalgamation or reproduction based on thousands of live footage videos, photographs, and sensory imaging. Weighing almost 10,000 kg in foam and fiberglass and held up by a single piece of steel pipe measuring only 16 inches wide, the whale's skeleton is, in fact, not whalebone but is instead an iron frame. Similar constructions of suspended whale bodies can be seen in the London Natural History Museum, in which a giant *Balaenoptera musculus* hangs on iron roof girders in a Victorian building, an installation that cost £12 million. Every year, the whale receives a full body scrub and vacuum session from museum staff on an electric lift.

Dioramas depicting environments had to be true to life if they were meant to capture nature in a fixed form. As German zoologist Benno Wandolleck explained in 1905, even the smallest mistake would render them comical.[2] To exploit the potential of dioramas fully, taxidermists and artists required extensive knowledge about the anatomy, behavior, and the "spirit" of the specimens. To grasp and capture the indefinable aspects of what taxidermy manuals referred to as "life expression" fully, the Department of Preparation, Arts, and Installation at the museum spent a considerable amount of time observing animals in their natural habitats.[3] Exhibit-makers frequently went on field trips with scientists to create field notes. In situations where field observation was not possible, curators would observe animals in zoos to study their muscular structures and movement patterns.

Natural history dioramas can be seen as one of the first institutionalized attempts to freeze past environments in time, bringing nature to the public eye. In the 19th and 20th centuries, natural history museums began to gain national attention as institutions dedicated to collecting specimens, cataloging extinct life, and educating a largely urban public about the "natural world." Richard Owen, the first superintendent of the Kensington Natural History Museum in London, imagined the space as a "cathedral to nature that anyone could visit for free."[4] Between 1869 and 1910, more than 30 natural history museums were built across North America, from the American Museum of Natural History in New York City (1869) to the National Museum of Natural History in Washington, DC (1910). At the turn of the 20th century, the director of the American Museum of Natural History, Morris K. Jesup, sent scientists and archaeologists on expeditions around the globe to collect traces of the extinct, the "exotic," and the disappearing foreign ecologies across the planet.[5] Brought back as preserved specimens in habitat dioramas in the museum, the "wild environments" featured in such displays included species that had once lived in niches far beyond the limits of Manhattan, showcasing plants and animals that one would never otherwise encounter in the city. Linked with Central Park, the Museum of Natural History is both an urban spectacle and a continuation of the nature–myth narrative that extends the urban park into the multisensory worlds of nonhuman life.

As historian Victoria Cain argues, exhibit makers and taxidermists of environmental dioramas gained esteem in the 1920s as technologists who *not only* possessed scientific knowledge about the natural world, but who were, in the public eye, also scientists who understood techniques of dissection, preservation, and staging to recreate scenes of wilderness in glass displays.[6] Whether through stuffing animal bodies, painting and embellishing patterns, or reassembling bones, the aim was that the animals and plants had to *look* like they were still alive, roaming in their native habitats with their companion species, captured mid-prowl. Over time, these technologies of environmental preservation would shift from taxidermy toward immersive mediums like augmented reality, mediating megafaunas of past and futures.

Back at the museum's Hall of Ocean Life, 750 other life-sized replicas are on display in staged scenes. In one diorama under the blue whale's massive tail, a glass scene of an ocean view displays an island, scaled down to appear as an archipelago far off in the distance. The sky is a pale cerulean blue, painted in mural form as a backdrop to a lively assemblage of kelp, mollusks, and sea urchins. The sky is faded in a gradient so that where white haze touches the water, two media appear to blend into one. Complete with the constant maintenance of imported sand and pebbles, the diorama is curated to reproduce the environmental time of its "inhabitants" continuously.

One year later, I traveled across the continent to Vancouver Island for fieldwork. At Canada's Royal BC Museum in Victoria, I meandered through an immersive exhibition with the uncanny title, *Hall of Advanced Mammals,* a tunnel-like space that led me from the Pliocene to the Holocene, into an open hall in which a massive woolly mammoth demands full attention. The diorama literally became the "elephant in the room" among smaller displays of now-extinct otters, ferns, gulls, and crustaceans. A tiny "extinct" label could be found next to the giant *Elephantidae* on the curatorial text. The exhibition contextualizes climate change into deep geological earth time, from the Ice Age to the Anthropocene.

I passed through many earth-times. At the end of the show, visitors are funneled into an immersive IMAX theater, replete with a 40-foot-high, 60-foot-wide screen. Most 3D theaters in natural history museums offer a similar selection of documentaries on deep ocean life, African megafauna, or some variation of Jurassic Park. At New York's American Museum of Natural History, a recent offering was a 39-minute film about the annual migration of giraffes, cheetahs, and elephants in the Serengeti preserve, a region in the north of Tanzania and extending to the southern border of Kenya. According to the description:

> More than one million wildebeest traverse this ecosystem every year, encountering lions, zebras, cheetahs, elephants, crocodiles, and many other remarkable animals along the way. Audiences will experience this amazing journey through the eyes of the youngest animals.

The moving image unfreezes frames into moving temporalities, projecting into the future while stringing together the past. Nevertheless, such environmental documentaries typically end with a moral call to action enveloped in human hubris–in this Anthropocene, human civilizations are "destroyers" of the otherwise pristine wilderness that must simultaneously retreat from nature reserves while coming to the rescue of less fortunate nonhuman species. In this framework, the fate of the nonhuman world depends solely on us. William Cronon's canonical text, "The Trouble with Wilderness: Or, Getting Back to the Wrong Nature," highlighted the fraught silos of the two spheres, arguing that preservation has historically conceived of "wilderness" as a last piece of untouched land "unspoiled by men."[7] The trouble with wilderness is that its purity both feeds into colonial discourse and prompts efforts to freeze or preserve nature in its most pristine state, either as dioramas or in the cabinets of natural history display. These dioramas perpetuate the idea of an ever disappearing, dwindling nature threatened by urbanization and driven by anxious attempts to save its last remaining pieces.

At the same time, the diorama display might be seen as a technology that hybridizes the natural and the technological, the organic and the inorganic. The American Natural History Museum, as a cultural institution and highly mediated space, produces life-size nature-simulacra of a pristine historical environment through realistic staging, multisensory affect, and taxidermic technologies. Seen as a technology, the diorama could be considered an early prototype of the virtual zoo, which later emerged with the advance of virtual reality (VR) technology in the 1990s.

Alien Zoo

On the second floor of the Westfield Century City Shopping Mall in Los Angeles, situated between Eataly and the international food court, visitors once had the chance to pay a $20.00 general admission fee to see one of several VR nature immersions run by the VR adventures company Dreamscape. Between savoring a gelato from Eataly's food court and a steak dinner at Del Frisco's, the average mallgoer could explore a prehistoric time, encounter dinosaurs, and communicate with deep ocean life all in one visit. While Dreamscape's US venture has transitioned into education VR tech since March 2024, adventure offerings once promised intimate encounters with charismatic megafauna in mist-filled rooms, where visitors were outfitted in multisensory gear that propelled them into the "sixth dimension."[8] Still open at the Geneva and Riyadh Park Mall locations, the official trailer states: "Here, everything changes. The water goes down to 0°C. The coral reef teems with vibrant sea life. Plunge into crystal-clear waters and come eye to eye with the largest animals on Earth—glide among enormous blue whales while helping reunite the blue whale family!"[9]

In an immersive experience called *THE BLU: Deep Rescue,* visitors become elite oceanographers submerged in the wonders of the deep blue ocean. Dreamscape's VR adventures delight visitors from across the world. Like the continuous halls of the Natural History Museum, which mixes temporal habitats from different geological eras into one experience, Riyadh's Dreamscape *Alien Zoo* VR Experience allows for the simultaneous co-existence of completely absurd animal-plant ecologies in an interstellar Jurassic Park environment.

1 American Museum of Natural History, "Blue Whale Model," https://www.amnh.org/exhibitions/permanent/ocean-life/blue-whale-model (accessed September 2024).

2 As quoted in Lynn Nyhart, "Die Kulturgeschichte des Zoos," *Isis* 95 (2004): 281–82.

3 See Oliver Davie, *Methods In The Art Of Taxidermy* (Hann & Adair 1894), 262.

4 The Natural History Museum, "A History of the Natural History Museum Grounds and Gardens," https://www.nhm.ac.uk/about-us/a-history-of-the-museum-grounds-and-wildlife-garden.html (accessed June 2024).

5 Plant expeditions sent scientists and technicians to Siberia, Congo, and Mongolia, to name a few.

6 Victoria Cain, "The Art of Authority: Exhibits, Exhibit-Makers, and the Contest for Scientific Status in the American Museum of Natural History, 1920–1940," *Science in Context* 24, no. 2 (2011): 215–38. Also, Karen A. Rader & Victoria E. M. Cain, *Life on Display: Revolutionizing U.S. Museums of Science and Natural History in the Twentieth Century* (University of Chicago Press, 2014).

7 William Cronon, "The Trouble with Wilderness: Or, Getting Back to the Wrong Nature," *Environmental History* 1, no. 1 (1996): 7–28.

8 The American company, Dreamscape Immersive, is an entertainment and technology enterprise specializing in immersive stage sets financially supported by major investors in Hollywood such as 21st Century Fox, Warner Bros, Nickelodeon, and AMC Entertainment. The "sixth dimension" is part of Dreamscape's branding to suggest a level of interactivity and immersion that feels beyond the typical five human senses or the fourth dimension of time. It could also refer to the simultaneous existence of more than one reality, each with its own timescale.

9 See THE BLU: Deep Rescue Trailer at YouTube, https://www.youtube.com watch?v=7JrvgT3lync (accessed September 2024).

10 See https://www.dreamscapemena.com/en-sa/adventures/details/alien-zoo.

11 For further reading on the notion of future megafaunas, see Jens-Christian Svenning, "Future Megafaunas: A Historical Perspective on the Scope for a Wilder Anthropocene," in Anna Tsing, et al. (eds), *Arts of Living on a Damaged Planet* (University of Minnesota Press, 2017), 67–86.

12 See Dreamscape Learn: https://www.dreamscapelearn.com/products.

13 Michael Sorkin, *Variations on a Theme Park: The New American City and the End of Public Space* (Hill and Wang, 1992).

14 Matthew Chrulew, "Reversing Extinction: Restoration and Resurrection in the Pleistocene Rewilding Projects," *Humanimalia*

In *Alien Zoo*, Pleistocene-era ferns and plants like *Picea glauca* (white spruce) thrive alongside imaginary, not-yet-existent future species such as *Astafie* skyray, a metallic blue fish-bird interspecies that rides on heat columns and airwaves, or *Murai* frog-cat, which, as the name suggests, is an amphibian-mammal genetically crossed with the common house cat. The official commercial for *Alien Zoo* reads:

> ALIEN ZOO. ...An adventure 1,000 light years in the making. SAVED FROM EXTINCTION. ...Volcanic eruptions, meteor impacts, and alien wars pose threats to endangered species across the galaxy. The intergalactic Wildlife Federation protects them in an orbiting wildlife refuge, where each creature feels right at home in its preternatural habitat.

At Riyadh Park Mall's *Alien Zoo*, visitors outfitted as conservationist-explorers come face to face with alien creatures–frolicking with frog-cats, soaring with the skyrays, and greeting the *Elqui*–"friendly creatures resembling horses but genetically related to manatees."[10] With the increase of digitized environments like Dreamscape Immersive, shopping malls and spaces of consumption depart from having real landscape elements in physical space. Managed by private companies and packaged into experiences, temporal access to digitized natures is contingent on an admission fee and a pre-scripted nature journey. Although the *Alien Zoo* is designed such that each encounter with creatures appears to be organic and completely randomized, it is a highly mediated and orchestrated affair, choreographed to play out in the exact same way with the same outcome with each turn.

As one Reddit user explained, "[I] got to touch one as it pressed up against my hand. ...I also threw a ball to some other little creatures." In the virtual zoo, "park visitors" can move unrestricted in a hybrid world within a virtual environment and manipulate physical objects by using motion capture technology in real-time, full-body mapping, virtual reality headsets, and physical sets that replicate actual surroundings. In this technological imaginary, the environment of the future exists beyond human civilization altogether, externalized onto another planet in an ambiguous space called the *Interstellar Wildlife Refuge*. At the same time, its format as an alien zoo implies human colonization in outer space, in which the term environment is expanded to encompass not just the unfamiliar but entirely unknown future fauna from other alien planets.[11] Curiously, Dreamscape LA has transitioned from mall entertainment to VR classroom focused on cinematic storytelling and pedagogy.[12] The *Alien Zoo* experience has been repackaged into a series of module-based educational "biology labs" for K–12 and university courses. From the mall to the classroom, biology students can visit *Alien Zoo* as field biologists. Teachers can transport their students to any environment in the universe at any point in history.

While the postmodern mall as *heterotopia* or as amusement park has been widely theorized, from Alexandra Lange in *Meet Me by the Fountain* to Michael Sorkin's *Variations on a Theme Park,* even Sorkin would not have dreamed of the hyper-simulacra of nature when he wrote of emerging urbanisms, "both at the heart and at the edge of town...megamalls, gentrified zones, and pseudo-historic marketplaces."[13] The heavily mediated environments offered by Dreamscape Immersion use technology to preserve a simulated wilderness in the middle of the city, in a world of rapid climate change, deforestation, and ongoing human-caused biodiversity loss and extinctions. While Dreamscape's *Alien Zoo* brings viewers into the spatiotemporal envelope of a virtual future, Sergey Zimov's notorious Pleistocene Park teleports the park visitor far back into the environment of a prehistoric past.

Pleistocene Park

Pleistocene Park, located in the northeastern portion of the Sakha Republic, is planned to populate more than 16,000 ha of land with gigantic northern mammals and megafauna that once lived in what is now Siberia. Russian scientist Sergey Zimov has been advocating for the reintroduction of keystone species like reindeer, bison, and Yakutian horses that have migrated farther north. To bring back Pleistocene environments and the "mammoth steppe" biome, melting tundra landscapes must be replaced with grassland, creating a suitable eventual home for the soon-to-be-resurrected woolly mammoth. Many journalists and critics have dismissed the plan (and de-extinction in general) as pseudo-scientific and utopian, comparing it to Jurassic Park or a theme park. The sheer size of Pleistocene mammals, they argue, could pose a risk to humans and livestock, could overtake and wipe out current habitats, and even transmit new disease strains while bringing back old viruses.

Although conservation biologists have attempted to restore ecosystems for decades, Pleistocene Park is completely novel in many ways, as its ambition is to recreate a grassland ecosystem from more than 100,000 years ago, taking restoration to a whole new prehistoric timescale. Zimov's rewilding plan flies in the face of conventional preservation, which traditionally takes the baseline "natural state of wilderness" to be the one encountered during the arrival of European colonization. The Pleistocene project defines "true wilderness" to be one that is not just pre-European but even pre-human. Yet, while Pleistocene Park expands the timeline of baseline wildernesses into the deep past, it continues to assume that there was one ultimate, singular, and original wilderness. As such, the park calls upon debates about nativity and what Matthew Chrulew problematizes as an obsession with "bioregional purity."[14] Which nature[s] and which kin do we claim as ours to resurrect? Do we revive everyone, or just the species that we find attractive? In *The Greening Imaginary,* sociologist Hilary Angelo argues that urban greening is a particular type of moral belief, a "social imaginary," in that it motivates action, is widely legible, and is a social form that can be filled with various substantive contents.[15] If we resurrect "native" species, then should we extend our attention to foreign species, including those often categorized as "alien," "invasive," and what Deborah Bird Rose once termed the "unloved others"?[16]

The biggest irony of the Pleistocene Park project is that while scientists are trying to restore a prehistoric environment, the actual means to achieve such an environment is highly mediated and requires heavy human intervention, including terraforming, predictive modeling, and computer simulations to test out various environmental conditions and population models. In one published image, Sergey Zimov drives a giant decommissioned military tank to mimic the stampede of a mammoth. What's more, de-extinction requires scientists to use cloning and DNA extraction technology. The attempt to resurrect the woolly mammoth by restoring its DNA has led people like Stephanie Turner to coin the term *genome time*, where we can now reverse the evolutionary process by seeing all extinct species as an endless stream of data and permutations subject to human control. Animals become reducible to code, in which the idea of extinction is no longer a problem, since the possession of a copy of genetic code means that any past environment could be, in theory, easily resurrected and restored.[17]

In such emergent technologies, environments become programmable and manageable by genetic components. Yet, consideration of which environmental time deserves preservation–and who does the resurrecting–is both a moral and political question. Natural history dioramas, alien zoos, and attempted revivals of the Pleistocene simultaneously cohere and defy the limits of bioregions and conventional environmental time. At a time gripped with climate anxiety and rapid extinction, VR zoos and ecological habitat simulations promise to offer a glimpse into a fantastical past–the illusion of a planet without human presence–and a fantastical future that hosts novel species as well as resurrecting the long extinct. New cyborgian landscapes and technologies challenge definitions of the natural world and illustrate our own multispecies entanglements in nature processes. As Latour states, "Those who wish to protect natural ecosystems learn, to their stupefaction, that they have to work harder and harder–that is, to intervene even more, at always greater levels of detail, with ever more subtle care–to keep them 'natural enough' for Nature-intoxicated tourists to remain happy."[18] The idea of environment, like the idea of nature before, is increasingly acknowledged as a Frankensteinian socially and culturally produced endeavor.

Whether embodied in cultural institutions in urban spaces (natural history exhibits), presented as VR experiences (commodity-centered nature), or manifested in prehistoric park reserves (de-extinction through genetic engineering), the category of environment is unstable and open to reinvention. As entry points, these environmental reproductions illustrate "ontologies of the present that demand archaeologies of the future."[19] Human intervention in the very process of evolution generates new entanglements and even new spatial–temporal arrangements in which human and animal life relationships are redefined, disrupting the idea of distinct ecological eras or environmental time. As humans begin to intervene in the past of evolution itself, we begin to have a direct hand in what gets a second life and what must remain dead.[20] As Irus Braverman writes in *Wild Life: The Institution of Nature,* species are essentially "prioritized according to their level of conservation reliance - namely, species who are not self-sustaining are privileged over others" for intense human management and interventionist stewardship in environmental management.[21] Our personal biases in the process of selecting candidates for de-extinction, for philosopher of science Derek D. Turner, transforms ecosystems into highly curated spaces based on the wishes of those who will fund resurrection technologies and synthetic biology conservation. The entire planet could, eerily, become our zoological garden.[22]

2, no. 2 (2011): 4–27. See also Matthew Chrulew, "Managing Love and Death at the Zoo: The Biopolitics of Endangered Species Preservation," *Australian Humanities Review* 50 (May 2011).

15 See Charles Taylor, *Modern Social Imaginaries* (Duke University Press, 2004); Cornelius Castoriadis, *World in Fragments: Writing on Politics, Society, Psychoanalysis, and the Imagination*, trans and ed. David Ames Curtis (Stanford University Press, 1997); Benedict Anderson, *Imagined Communities: Reflections on the Origin and Spread of Nationalism* (London: Verso, 1983).

16 See Deborah Bird Rose & Thom van Dooren (eds), *Unloved Others: Death of the Disregarded in the Time of Extinctions*, theme issue *Australian Humanities Review* 50 (May 2011).

17 Stephanie S. Turner, "Open-Ended Stories: Extinction Narratives in Genome Time," *Literature and Medicine* 26, no. 1 (2007): 55–82.

18 Bruno Latour, "Love Your Monsters: Why We Must Care for Our Technologies as We Do Our Children," *Breakthrough Journal* 2 (2011): 24.

19 Fredric Jameson, *A Singular Modernity: Essay on the Ontology of the Present* (Verso, 2002), 215.

20 See Foucauldian concepts of biopower: "[A] power that exerts a positive influence on life, that endeavors to administer, optimize, and multiply it, subjecting it to precise controls and comprehensive regulations." As seen in *The Will to Knowledge: The History of Sexuality Volume 1* (Pantheon Books, 1978), 137.

21 Irus Braverman, *Wild Life: The Institution of Nature* (Stanford University Press, 2015), 223.

22 For more debate on de-extinction, see Yale Environment 360, "De-Extinction Debate: Should We Bring Back the Woolly Mammoth?" https://e360.yale.edu/features/the_case_for_de-extinction_why_we_should_bring_back_the_woolly_mammoth (accessed September 2024).

LAURA J. MARTIN

BEYOND HABITAT

Laura J. Martin is an associate professor of environmental studies at Williams College. Her research on biodiversity has been featured in venues including the *New York Times* and the *Atlantic*, and was recently recognized by the Falling Walls Foundation as breaking the barriers between the sciences and the humanities. She is the author of *Wild by Design: The Rise of Ecological Restoration* (Harvard University Press).

+ ECOLOGY, ENVIRONMENTAL HISTORY

Climate change and persistent pollutants threaten to make habitat uninhabitable. What happens then? When we look to a future with less available habitat, we imagine suffering (mass death, forced migration, and societal collapse), epic escape (technological miracles, enclaved utopias, and colonizing the moon), or dismal compromise (such as Tuvalu's plans for "digital migration").[1] The Intergovernmental Panel on Climate Change warns of permanent habitat loss due to flooding, sea level rise, and erosion.[2] Recent news headlines include "23% of Earth's Natural Habitats Could Be Gone by 2100," "Climate Change and Habitat Loss Push Amphibians Closer to Extinction," and "Animals are Running Out of Places to Live."[3]

Since ideas of environmental threat and environmental futures are so often about habitat, we must question what habitat is, where it came from as an idea, and most importantly, how we might think beyond habitat. This is not just a theoretical exercise but a crucial step toward a more sustainable and inclusive future.

The word *habitat* emerged from the discipline of natural history, and its first known use was in W. Withering's *Arrangement of British Plants* (1796) as "the natural place of growth of a plant in its wild state."[4] *Habitat* appeared sporadically in 19th-century European biogeographical literature, although Darwin did not think in terms of habitat. In *On the Origin of Species*, Darwin explained speciation as the result of selection for variations "in any degree profitable to an individual of any species, in its infinitely complex relations to other organic beings and to external nature" – a much more extensive rendering of selective forces than habitat.[5]

In the early 20th century, however, ecologists organized their new scientific discipline around the concept of habitat. In 1905, English botanist Arthur G. Tansley defined ecology as the study of "those relations which depend directly upon differences of habitat among plants."[6] Early ecologists worked to describe and differentiate habitats by their "life conditions" (environmental variables), which they believed drove selection. As Tansley put it: "We find the plants of the mountain differing from those of the valley, the plants of marshy land differing from those of the dry plain, the plants of the coast differing from those of the interior, the plants of the sand-dunes differing from those of the salt-marshes."[7] Ecology was the science of explaining plant forms shared across vast geographic differences; why, say, montane plant species in Switzerland looked like species in Colorado. Ecologists contended that habitat determined form.

Habitat remains a foundational concept in ecology, and ecologists have used it in two ways: to refer to a geographic space (the area a species occupies) or to a conceptual space (a set of biophysical conditions necessary for a species' survival).[8] Both senses have structured experiments, theories, and legal frameworks like the US Endangered Species Act.[9] Just as importantly, the two meanings of habitat have created intellectual and political impasses for evolutionary ecologists and environmentalists.

Intellectually, the idea that species are adapted to habitat fails to explain the success of invasive non-native species. If species evolve to fit their habitats, what are we to make of species that thrive everywhere?[10] Further, what is a "natural place of growth" when people have been moving and cultivating organisms for millennia? Separately, the idea that species need particular biophysical conditions to survive fails to account for those species that directly modify their habitats. The beaver dam and the superhighway alike suggest that habitats are not rigid molds that shape species because species, in turn, remake those molds. This insight underpins niche construction theory, which emphasizes the capacity of organisms to modify environments and hence evolution, and it motivated me and my collaborators to develop the field of indoor biome studies.[11]

In creating habitat for people (and pets and other domesticated animals), designers and builders have unintentionally created habitat for myriad other species. The indoors is a massive habitat found on every continent. Thousands of species–likely hundreds of thousands–live in the indoor biome, many preferentially or obligately. One study of 50 houses in North Carolina, US, found more than 500 species of arthropod.[12] My collaborators and I found that the indoor biome is substantial globally, spanning an area the size of France and similar in size to the flooded grassland biome and the tropical coniferous forest biome.[13] Unlike those biomes, the indoor biome is expanding, yet urban ecology research is conducted almost entirely outdoors.[14] This discrepancy is a problem for anyone seeking to understand the impact of people on other species. For instance, the indoor biome of Manhattan is nearly three times larger in terms of floor space than the island's area. Like "invasive" non-native species that seemingly defy the confines of habitat, niche construction theory challenges efforts to characterize habitat, which is continuously being reshaped.

Politically, the idea of habitat reaches a different impasse through the limited possibilities of habitat preservation. The preservation movement is often said to have begun in 1872 when the US federal government established Yellowstone National Park by forcibly dispossessing Crow, Shoshone, and Bannock communities of their land and resources.[15] Notably, protecting the habitat of nonhuman species was not the goal of early preservationists; instead, they hoped to provide beautiful scenes to settler tourists and revenue to railroad companies. Calls to preserve habitat for biodiversity did not gain traction until the 1960s.[16] Since then, the number of protected areas worldwide has skyrocketed, as has their extent.[17] This approach has led to the consolidation of control of habitat by nation-states, often at the expense of local and Indigenous communities. It is past time for environmentalists and academics to reject this approach.

1 "The First Digital Nation," https://www.tuvalu.tv/ (accessed July 3, 2024).

2 Intergovernmental Panel on Climate Change (IPCC), *Climate Change and Land: An IPCC Special Report on Climate Change, Desertification, Land Degradation, Sustainable Land Management, Food Security, and Greenhouse Gas Fluxes in Terrestrial Ecosystems* (IPCC, 2019), 148, 358, https://www.ipcc.ch/site/assets/uploads/2019/11/SRCCL-Full-Report-Compiled-191128.pdf. See also IPCC, *Climate Change 2022: Impacts, Adaptation and Vulnerability*. Contribution of Working Group II to the Sixth Assessment Report of the Intergovernmental Panel on Climate Change (Cambridge University Press, 2022).

3 "Climate Change and Habitat Loss Push Amphibians Closer to Extinction," *Nature* (October 4, 2023), https://www.nature.com/articles/d41586-023-02785-1; Carly Nairn, "23% of Earth's Natural Habitats Could Be Gone by 2100, Study Finds," *World Economic Forum* (November 12, 2020), https://www.weforum.org/agenda/2020/11/earth-natural-habitats-destroyed-biodiversity-loss/; Catrin Einhorn & Lauren Leatherby, "Animals are Running out of Places to Live," *New York Times* (December 9, 2022), https://www.nytimes.com/interactive/2022/12/09/climate/biodiversity-habitat-loss-climate.html.

4 Oxford English Dictionary, "Habitat," https://www.oed.com/dictionary/habitat_n?tab=factsheet#2153169 (accessed July 1, 2024).

5 Charles Darwin, *On the Origin of Species by Means of Natural Selection* (D. Appleton and Company, 1861), 61.

6 A. G. Tansley, "The Problems of Ecology," *The New Phytologist* 3, no. 8 (1904): 191–200.

7 Ibid., 194.

8 Peter S. Alagona, "What is Habitat?," *Environmental History* 16, no. 3 (2011): 391–97.

9 On habitat and law, see Alagona, *After the Grizzly: Endangered Species and the Politics of Place in California* (University of California Press, 2013); Laura J. Martin, *Wild by Design: The Rise of Ecological Restoration* (Harvard University Press, 2022), Chapter 6.

10 I take up this question in *Wild by Design*.

11 K. Laland, B. Matthews & M. W. Feldman, "An Introduction to Niche Construction Theory," *Evolutionary Ecology* 30 (2016): 191–202; Laura J. Martin, et al., "Evolution of the Indoor Biome," *Trends in Ecology & Evolution* 30, no. 4 (2015): 223–32.

12 Matthew Bertone, et al., "Arthropods of the Great Indoors: Characterizing Diversity Inside Urban and Suburban Homes," *PeerJ* 4:e1582, https://doi.org/10.7717/peerj.1582.

13 Martin, "Evolution of the Indoor Biome," 223–32.

Preservation is the assertion that governments should protect nonhuman habitats from human habitation, and today, preservation dominates environmental policy and action. In 1900, there were 164 protected areas worldwide, encompassing about 31,000 km2, which is the size of Vancouver Island, Canada. By 1960, 9,289 protected areas encompassed roughly 1.4 million km2, slightly larger than Peru. As of July 2024, 294,035 protected areas encompassed 50.9 million km2, an area larger than the continent of Asia. This land and water is controlled by nation-states, not by communities.[18] Before 1900, national and subnational government agencies were responsible for making decisions about 74% of protected areas, followed by nonprofit organizations at 17%. As of 2022, governments manage 92% of protected areas, followed by nonprofit organizations at 2.7%, individual landowners at 3.6%, and local and Indigenous communities at just 0.3%. As a consequence of the preservation movement, nation-states have consolidated control of habitat.

Enthusiasm for habitat preservation is increasing. Approximately 17% of land and inland waters, and about 8% of marine areas, are currently protected.[19] The recent Global Deal for Nature aims for 30% of Earth to be formally protected and an additional 20% to be designated as "climate stabilization areas" that would offset greenhouse gas emissions. The proposal enjoys strong support among scientists, diplomats, and environmental organizations. E.O. Wilson endorsed this 50% target in his 2016 book *Half-Earth: Our Planet's Fight for Life*, writing that "only by setting aside half the planet in reserve, or more, can we save the living part of the environment."[20] Social scientists estimate that if Wilson's half-Earth plan is implemented, more than one billion people could become conservation refugees, banished from their homes and livelihoods in the name of protecting other species.[21] The expansion of protection areas will exacerbate inequalities in the places that have historically contributed least to climate change and biodiversity loss. Is this how we want to live with others – with other people and with other species?

The verb *to inhabit* can offer us more, intellectually and politically, than the noun *habitat*. It bears noting that the verb *inhabit* is much older than *habitat*; people and other species *inhabit* at least as far back as Chaucer's 1374 translation of Boethius's *The Consolation of Philosophy*.[22] To inhabit is an action, whereas habitat is a thing. Habitat depends upon static claims to belonging, ownership, and futurity, whereas inhabiting is a dynamic process.[23]

We would do well to dwell with the verb form, which brings us most directly to the question: How do we want to inhabit? What is the ethical way to live with each other and with other species? How will we share our most important and cherished places? And can we recognize that we are not the only species who inhabit; who come to know and love and depend upon places, and who also depend upon places we do not know and love?

Environmental injustice and biodiversity loss are often viewed as unrelated issues, but both result from habitat-thinking: the presumption that the world can be divided into discrete parcels of inhabited and uninhabited land. Preservation is segregation.

14 Laura J. Martin, Bernd Blossey & Erle C. Ellis, "Mapping Where Ecologists Work: Biases in the Global Distribution of Terrestrial Ecological Observations," *Frontiers in Ecology and the Environment* 10 (2012): 195–201.

15 Mark Spence, *Dispossessing the Wilderness: Indian Removal and the Making of the National Parks* (Oxford University Press, 2000).

16 Martin, *Wild by Design*, Chapter 7.

17 The data for this figure comes from UNEP-WCMC and IUCN (2024), "Protected Planet: The World Database on Protected Areas (WDPA)" and "World Database on Other Effective Area-based Conservation Measures (WD-OECM)," www.protectedplanet.net (accessed July 23, 2024). The figure excludes sites where the legal status is "Proposed" or "Not Reported."

18 The data for this figure comes from UNEP-WCMC and IUCN (2024). "Government" includes the IUCN governance types "federal or national ministry or agency," "subnational ministry or agency," "government-delegated management," and "transboundary management." "Other" includes the IUCN governance types "collective governance," "joint governance," and "for-profit organizations." The figure excludes protected areas with a governance type "not reported."

19 Protected Planet, "Discover the World's Protected and Conserved Areas," https://www.protectedplanet.net/en (accessed July 23, 2024).

20 E. O. Wilson, *Half-Earth: Our Planet's Fight for Life* (Liveright, 2017), 3.

21 Judith Schleicher, et al., "Protecting Half of the Planet Could Directly Affect Over One Billion People," *Nature Sustainability* 2 (2019): 1094–96; Mark Dowie, *Conservation Refugees: The Hundred-Year Conflict between Global Conservation and Native Peoples* (MIT Press, 2011); Laura J. Martin, "Avoiding Carbon Colonialism," *The Hill* (April 26, 2021), https://thehill.com/opinion/energy-environment/550313-avoiding-carbon-colonialism/.

22 Oxford English Dictionary, "Inhabit," https://www.oed.com/dictionary/inhabit_v?tab=meaning_and_use#449334 (accessed July 1, 2024).

23 I thank the faculty and students of the 18th Ischia Summer School on the History of the Life Sciences, "Humans, Natures and the Nature of Humans," for inspiring this essay. Our jumping-off point was Mauricio Onetto Pavez, "Habitability as a Historical Category for Interpreting the Anthropocene," *Journal of Historical Geography* 83 (2024): 96–109. On early modern understandings of the planet and mobility, see Suman Seth, "Race, Seasoning, and Strangeness: The Problem of Mobility in Early Modern Science and Medicine," *Global Intellectual History* 8, no. 4 (2022): 448–66.

Preservationists assert that people cannot inhabit the world without destroying the habitat of other species – that people and other species cannot cohabit. Simultaneously, preservation is a form of political resignation; it concedes that selfish people, corporations, and systems will continue to harm other species without abatement. Preservationists believe in the undifferentiated *Anthropos* of the Anthropocene, in the view that *Homo sapiens* is a destructive species that needs its own habitat apart from the habitat of other beings. The preservation movement and initiatives like the Global New Deal promise to divide the world into sacrifice zones and protected zones, into habitat for people and habitat for everything else. This way of thinking—and of acting—begets the two terrible problems of our times: the poisoning of people, especially impoverished people and people of color, and the loss of biodiversity.

Habitat implies that the environment comprises many nested habitats, each species with its bespoke requirements. Inhabiting gestures to something more significant. It is true that life on Earth is diverse; perhaps our human world shares little in common with the world of a bacterium that lives on sulfur rather than oxygen. But many beings do share worlds. There will always be a scale at which we cohabit. For most of us, that scale is not the planet. It is the home.

Design is a mode of inhabiting, as is ecological restoration. In *Wild by Design*, I define ecological restoration as an attempt to co-design nature with nonhuman participants.[24] Unlike hands-off preservation, restoration is an active intervention. Restorationists aim to undo human-caused harm to species or ecosystems. However, restoration is not an unrestrained intervention: restorationists strive to cede some control of the restoration process to other organisms. And unlike interventionist natural resources management, which aims to sustain the economic exploitation of particular species into the future, restoration focuses on unharvested species, valued in terms that exceed and defy economics. In its best form, ecological restoration operates as a form of care for other species and a mode of reconciliation with the human past.

Interest in restoration is growing worldwide. Public and private organizations spend billions of dollars annually on species reintroductions, invasive species removal, wetland creation, tree planting for carbon offsetting, and other restoration actions. Meanwhile, a suite of international agreements promote restoration, including the Ramsar Convention on Wetlands, the UN Framework Convention on Climate Change, and the Kunming-Montreal Global Biodiversity Framework. Recognizing restoration's potential—and its centrality to international environmental negotiations—the UN General Assembly recently declared 2021–2030 the UN Decade on Ecosystem Restoration. In 2024, the European Union passed a law requiring countries to restore nature on at least 20% of their lands and seas by the decade's end.

The field of restoration ecology is not organized around habitat, although the concept certainly influences some restoration work. The ecological restoration movement began with the work of ecologists and landscape architects in the 1920s. These restorationists had diverse goals, including saving particular plant and animal species, designing beautiful landscapes, fostering civic pride, and teaching future generations.[25] Initially, it was the purview of loosely organized networks of volunteers; today, restoration is increasingly privatized and consolidated. Large-scale restoration has some benefits; such projects can incorporate more species and may be more cost-effective than small-scale, local restoration.[26] However, large-scale restoration projects risk undermining democratic and local processes, and many areas identified by scientists and policymakers as high-priority for restoration are currently used for livestock or crops.[27] A large-scale restoration project is a large-scale imposition of just one set of aesthetics, values, and governance norms. Consolidation was hastened by the invention in the 1980s of "off-site" mitigation, the practice of compensating for ecological harm in one location with restoration at another. Unlike other modes of restoration, off-site mitigation is premised on habitat, specifically the fungibility of habitat. Off-site mitigation uncoupled sites of harm from sites of care, paving the way—conceptually and administratively—for today's carbon offset market. Off-site mitigation has redistributed ecosystems worldwide. It also has reconfigured relations between the global North and the global South, fashioning the latter as a source of "wild" habitat meant to compensate for pollution in the former.[28]

A second arena in which habitat enters restoration ecology is assisted migration. The current debate over the ethics of assisted migration divides restoration ecologists who prioritize geographic habitat from those who prioritize conceptual habitat. Atomic waste and legacy pollutants (e.g., PCBs and PFAS) are now understood to persist in the environment for tens of thousands of years. Meanwhile, scientists predict that climatic changes caused by increases in atmospheric CO2

will be largely irreversible for at least one thousand years after emissions stop. Species are no longer found where they used to be; shifts in temperature and rainfall patterns have led some species poleward or to higher elevations. Such geographic shifts could result in the uncanny scenario in which Joshua Trees only survive outside Joshua Tree National Park.[29] While migratory species can sometimes find new territories that suit them, other species face substantial barriers to movement: suitable habitat might be hundreds of miles away or across a city; further, species that depend on each other may not shift their ranges at the same rate, disrupting ecological interactions. Thus, prioritizing conceptual habitat, some ecologists argue that people should move species, or even whole communities of species, to new places where they could thrive in future climatic conditions. Proponents of this mode of restoration call it "assisted migration"; opponents, prioritizing geographic habitat and worrying about unintended species invasions, call it "assisted colonization" or "playing God."[30]

Landscape architects, designers, and restoration ecologists are well-positioned to think beyond the conceptual and political traps of habitat. These fields offer foundational concepts like beauty, continuity, resilience, flows, community, and experiment. By juxtaposing ideas and practices from these disciplines, we can tackle new questions about the value of historical fidelity in ecosystems and design, the relationship between the local, the regional, and the global, and the (mis)alignment of biodiversity goals and climate goals.

Restoration takes time.[31] It is an ongoing process of care and living together. If restoration fails to identify and stop ongoing environmental harms, it will have amounted to no more than habitat creation. If it succeeds, to restore will mean to inhabit.

24 Martin, *Wild by Design*, 5.

25 Ibid., Chapter 3.

26 Thomas M. Neeson, et al., "Enhancing Ecosystem Restoration Efficiency through Spatial and Temporal Coordination," *Proceedings of the National Academy of Sciences* 112, no. 9 (2015): 6236–41; Heida L. Diefenderfer, et al., "Applying Cumulative Effects to Strategically Advance Large-scale Ecosystem Restoration," *Frontiers in Ecology and the Environment* 19, no. 2 (2021): 108–17.

27 Forrest Fleischman, et al., "Restoration Prioritization Must be Informed by Marginalized People," *Nature* 607 (2022): E5–E6.

28 Ibid., Chapter 7.

29 See, for example, Camille Parmesan & Gary Yohe, "A Globally Coherent Fingerprint of Climate Change Impacts Across Natural Systems," *Nature* 421 (2003): 37–42; Rachael Hickling, et al., "The Distributions of a Wide Range of Taxonomic Groups are Expanding Polewards," *Global Change Biology* 12 (2006): 450–55; Philip Kiefer, "Iconic Joshua Trees May Disappear – But Scientists Are Fighting Back," *National Geographic* (October 15, 2018).

30 N. Hewitt, et al., "Taking Stock of the Assisted Migration Debate," *Biological Conservation* 144, no. 11 (2011): 2560-72; Laura J. Martin, et al., "Conservation Opportunities Across the World's Anthromes," *Diversity and Distributions* 20, no. 7 (2014): 745–55; Janet Marinelli, "Plants on the Move," *LA+ Interdisciplinary Journal of Landscape Architecture* 19 (2024): 34–39.

31 Laura J. Martin, "Slow Restoration, Rewilding, and Design," *Ri-Vista: Research for Landscape Architecture* (forthcoming).

IMAGE CREDITS

Endpapers

Keelmen Heaving in Coals by Moonlight, by J.M.W. Turner, 1835. Public domain via The National Gallery of Art.

Editorial

p. 4: Algal bloom up close, by Ian Jacobs, used under CC-BY-NC 2.0.

Environments, Environmentalisms, Envirotechniques

p. 6: Image by Shuying Rong, used with permission.

The Other Silence in *Silent Spring*

p. 12: Drawing by Megan Singleton, used with permission.

p. 15: Walter Reuther leads Farm Workers, Delano Rally, CA circa 1965–66. George Ballis / Take Stock / TopFoto.

p. 18: Drawing by Megan Singleton, used with permission.

In Conversation with Jess Conard

p. 20: Image by Shuying Rong, used with permission.

p. 25: Video stills from ABC 7 Chicago, https://www.youtube.com/watch?v=dsw4ZKkLQp8&list=WL&index=3 and NBC News https://www.youtube.com/watch?v=S1YZ7FYH-gk.

In Conversation with Maria Ivanova

p. 28: Image by Colin Curley, used with permission.

On the Road with Herbert Bayer

p. 34: Herbert Bayer, ARCO Refinery, Philadelphia, 1972, Red Pyramid, Photocollage, and Yellow Undulating Wall, Photocollage. Images © 2025 Artists Rights Society (ARS), New York / VG Bild-Kunst, Bonn.

p. 38: Herbert Bayer, ARCO Refinery, Philadelphia, 1973 sculptural screens (proposal). Photocollage. Denver Art Museum: Gift of BP Corporation, 2001.565. Image courtesy Denver Art Museum. © 2025 Artists Rights Society (ARS), New York / VG Bild-Kunst, Bonn; Herbert Bayer, Highway Gate in Three Parts, 1971 acrylic, collage, and graphite. Images © 2025 Artists Rights Society (ARS), New York / VG Bild-Kunst, Bonn.

p. 39: Clockwise from top left: Herbert Bayer, ARCO Refinery, Philadelphia, 1972 fence and building (proposal). Colored pencil drawing. Denver Art Museum: Gift of BP Corporation, 2001.569. Image courtesy Denver Art Museum; Herbert Bayer, Road in deep embankments, 1972 acrylic and charcoal; Herbert Bayer, Positive negative highway, 1971, wood, mat board, paint, toy automobile; Herbert Bayer, Highway landscape with four hills, 1972 pastel and collage. All images © 2025 Artists Rights Society (ARS), New York / VG Bild-Kunst, Bonn.

Sounding Extreme Environments

p. 42-43: Image by Luke Mackin, used under CC BY-NC-SA 2.0.

p. 48: Image by United States Federal Aviation Administration and Beranek Bolt. Public domain via the Library of Congress, https://www.loc.gov/item/gm69003433/.

Nuclear Weapons and the Unity of the Environment

p. 52-53: Photo courtesy of National Nuclear Security Administration / Nevada Field Office.

p. 56-57: Photos by Nick Carr, Scouting NY, used under CC BY-NC-ND 2.0.

p. 61: Manhattan Project Los Alamos main gate (cropped). Public domain via National Park Service.

In Conversation with Michael Oppenheimer

p. 62: Image by Qiyu Wang, used with permission. Photo of Michael Oppenheimer © World Economic Forum / Sikarin Thanachaiary, used under CC BY-NC-SA 2.0.

p. 66: Image by Qiyu Wang, used with permission. Memoradum is a copy of the original document from the Nixon Library, https://www.nixonlibrary.gov/sites/default/files/virtuallibrary/documents/jul10/56.pdf.

Spiritualizing the Two Cultures, c. 1978

p. 70: Image by Shuying Rong, used with permission.

p. 74: Buckminster Fuller's Scheme, Cathedral of St. John of the Divine Environmental Program. Redrawn by Shuying Rong.

Failure to Build

p. 78: Image by Megan Singleton, used with permission.

In Conversation with Dipesh Chakrabarty

p. 84, 86, 87: Collage by Colin Curley, used with permission.

Why Environmental History Matters

p. 90-91: Photos from Batik Art Workshop for the Tiger Widows' Children at Rajat Jubilee village in West Bengal, January 2025. Photographs by Anindita Saha, used with permission.

Snow Globes: Cyclinc Drawing and Cryospheric Panoramas

p. 102: "Pack-ice in the Ross Sea. Midnight January 1911," watercolor by Edward Wilson from Apsley Cherry-Garrard, *The Worst Journey in the World* Vol. 1 (1922). Public domain.

p. 106: Image courtesy of the British Museum.

p. 107: Section of the Rotunda, Leicester Square, exhibiting a panorama of London from the River Thames by Robert Mitchell (1801). Public domain, cropped with a circular frame.

p. 108: Image by Edward Wilson from Apsley Cherry-Garrard, The Worst Journey in the World, Vol. 1 (1922), public domain.

Megafaunas of the Past and Future

p. 110: Collage by Qiyu Wang, used with permission.

Beyond Habitat

p. 118: Photograph by Tom Lynn, used with permission

IN THE NEXT ISSUE OF LA+

Our knowledge of environments and landscapes is fused with the media we use to understand them, just as our environments and landscapes are suffused with media. Media technologies have significant environmental impacts, including energy consumption, electronic waste, and resource extraction. On the other hand, narrative forms of media play a pivotal role in environmental advocacy and education about environmental issues. Most importantly, media are in the middle; they are not neutral channels that merely transmit content created through other means. In our upcoming issue, **LA+ MEDIA** contributors will explore how various forms of media influence our knowledge, shape our understanding, and affect the way we interpret the world around us.

BROOKE BELISLE
ALENDA Y. CHANG
DAVE COLANGELO
DANIEL COOMBES
RANIA GHOSN
ABELARDO GIL-FOURNIER
ALISON GRIFFITHS
MARTIN HOGUE
BRYAN NORTON
JUSSI PARIKKA
JOHN DURHAM PETERS
ANN REYNOLDS
DAVID SALOMON
SUSAN SCHUPPLI
NINA LAGER VESTBERG

OUT SPRING 2026